# LEAN SIX SIGMA

## 3 BOOKS IN 1

The Complete Guide to Implementing Six Sigma
Methodology for Continuous Manufacturing Process
Improvement

## JEFFREY RIES

2

# BOOKS INCLUDED:

Lean Startup

Lean Analytics

Lean Enterprise

# Table of Contents

## Lean Startup

# Lean Analytics

# Lean Enterprise

# Lean Startup

*The Complete Step-by-Step
Lean Six Sigma Startup Guide*

# Introduction

Congratulations on getting a copy of *Lean Startup: The Complete Step-by-Step Lean Six Sigma Startup Guide* and thank you for doing so. There are two questions that any company can ask to both reduce unnecessary failure while at the same time ensuring that the company focuses only on ideas that have promising potential. They are:

- Should we build this new service or product?
- How can we improve our odds of success with this new thing?

The Lean method is equally useful for startup companies as it is for Fortune 500 companies. It may have its roots in the technology sector but it is already being used in virtually every industry across the board. While there is lots of confusion around it, the Lean Startup system can help companies of all sizes in a lot of different ways.

While the term "startup" generally has very specific connotations in the business world, in this instance, "startup" simply means any team that is planning to create a new product or service whose future isn't 100 percent certain yet. Generally speaking, it makes far more sense to classify startups as enterprises taking on the challenge amidst uncertainty, than by categories like market sector, size or even age of the company.

With this definition in mind, you will find that there are a few main areas in which a startup faces the greatest amount of uncertainty, otherwise known as risk. Technical or product risk can be summed up by the question "Can it be built?" As an example, doctors who are currently working towards a cure for cancer can be thought of as a startup institution because there is a very large technical risk and this area of study has been going on for quite some time with no hint of success. However, if they do discover a cure, there is absolutely no market risk because its target market would definitely buy it.

Market risk, also known as customer risk, is simply the risk when the product or service reaches the market and no one is actually going to want to buy it. A cautionary example of this type of risk is a company named Webvan that spent millions and millions of dollars creating an automated means of buying groceries online. The only problem is that they tried to get this system up and running in the early 2000s. This is a time when many people were still getting comfortable with the concept of the internet in general but the comfort in buying everyday products online did not follow until nearly a decade.

The business model risk is the risk associated with taking a good idea and building a functioning business plan around it. Even if you already have a good idea, the right business model could very well not be visible until the service is up and running. As an example, when Google started its original business plan of selling advertisements based on previous searches, the plan wasn't clear because no one had done that sort of thing before.

While every company will need to deal with these risks to varying degrees, the biggest risk that most new products or services struggle with is customer risk. It can be difficult to determine the value of something new for customers who haven't experienced it yet. The tricky part here is that in most instances, it will actually appear that the product risk is the most urgent risk. After all, most new ideas don't make it this far without an assumption that someone, somewhere is going to want the product or service at hand. This assumption, then, can lead to a much costlier course of action wherein you do the work to create the product or service before offering it to anyone.

This is where the Lean Startup system comes into play. This technology potentially stops you from being one of the millions of companies out there that has a good idea and a cool product but had crashed and burned because they inherently relied on assumptions about consumer behavior that simply turned out not be true. It is important to think in terms of risk as opposed to company history because in doing so, you will find that many large companies have startup organizations within them. As an example, consider the Gillette razor company who felt that there was little risk in adding the fifth blade to their flagship line of razors because they knew the business model, the market, and the product ins and outs. However, the company that owns Gillette, Proctor and Gamble, operates a startup in the form of its research and development division that focuses specifically on hair removal. With each new idea, this division seems like a startup because they have no known variable which means everything they are working on is extremely risky.

Currently, one of the well-known companies that using the Lean Startup system is General Electric, which is also one of the largest companies in the world. The company has trained more than 10,000 managers around the world to use Lean Startup principles and has used the system to successfully improve the end result on all of their products including refrigerators and diesel engines.

To follow in their footsteps, the following chapters will discuss how to operate a Lean Startup successfully, starting with an overview of the Lean Startup methodology. Next, you will learn how to create a trial startup system that is not only useful but also designed to provide you with as much viable information as possible. You will then learn how to take a successful startup and grow it until it reaches its full potential. From there you will learn about adding Six Sigma and other Lean tools to your startup for maximum efficacy.

There are plenty of books on this subject on the market, thanks again for choosing this one! Every effort was made to ensure it is full of as much useful information as possible, please enjoy!

# Chapter 1: Lean Startup Options

While the idea of the Lean Startup has been around since 2011, many companies are still coming to grips with everything the system has to offer. This is despite the fact that most of the ideas presented in this system were hardly new. This is largely due to the fact that the system actually offers more value to established organizations than it does to startups. However, startups can still be able to build a Lean system from the ground up if they choose to.

## Lean Startup methodology

*Build, measure, and learn:* Perhaps more than anything else in recent history, the application of the scientific method to demolish uncertainty, where innovation is concerned, has transformed the way breakthroughs happen. Broken down, this includes the process of defining a hypothesis, creating a prototype to test the hypothesis, testing the prototype (and thus the hypothesis) and then adjusting as needed. While this may seem simple, it has the potential to generate massive results by enabling companies to take risks on smaller ideas without breaking the bank in the process.

The build, measure, and learn approach can be used for virtually everything, not just entirely new ideas. It can be used to test things like customer service ideas, the process of managerial review, or even a new feature for an existing product or service. As long as you can perform a test that clearly validates or disproves the initial hypothesis, then you will be good to go because you must be able to gather enough data to justify approving or vetoing the idea.

The goal, then, is to do everything possible in order to ensure that build, measure, and learn process proceeds from start to finish as quickly as possible. This will make it feasible to run the process multiple times if needed, while also making it clear when such additional runs are needed. As such, it is important to have a very specific idea for each test because as more variables are added, the more difficult it will be to determine results with any real degree of accuracy. When it comes to products and services, this means determining if they are either wanted or needed by the target audience.

*Minimal viable product:* Generally speaking, most product development involves an extreme amount of work up front. The process involves working through the full specifications of the product, as well as a significant initial investment when it comes to capital in order to build and test multiple iterations of the product. The Lean Startup process thus encourages building only enough of the product in question to make it through a single round of the build, measure, and learn process at a time. This is what is known as the minimal variable product.

The minimal variation of the product is what enables a full cycle of the build, measure, and learn loop to be completed with the least amount of required time and effort on the part of the team. This may not be something as simple as writing a new line of code, it could be an elaborate process that outlines the customer journey, or a complete set of mockups made out of a cheaper substitute. As long as it is enough to test the hypothesis, then it is good to go.

*Validated Learning:* An important part of the Lean Startup

process is ensuring that you are testing your hypothesis with an eye towards the right metrics. Failing to do so can make it easy to focus on vanity metrics instead. Focusing on vanity metrics may make you feel as though you are making progress while not actually telling you all that much about the value of the product. For example, for Facebook, the vanity metrics are the things like the total number of "Likes" that have been received or the number of total accounts created. The real meat and potatoes are in metrics such as the amount of time the average user spends on the service per week. Early on, the metric that validated the company's initial hypothesis was the fact that more than half its user base came back to the service every single day.

*Innovation accounting:* Innovation accounting is what makes it possible for startups of all sizes to prove, in an objective way, that they are creating a sustainable business. The process includes three steps, starting with determining the baseline. This involves taking the minimum viable product and doing what you can to determine relevant datapoints that can be referred back to the fact. This could involve things like a pure marketing test to determine if there is actually interest from customers. This, in turn, will make it possible for you to determine a baseline with which to compare the initial cycle of the build, measure, and learn process, too. While better numbers are always desired, poor numbers at this stage aren't terribly important, it only means that the team will have more work to do in the build, measure, and learn cycle.

After the baseline has been determined, the next step is going to be to make the first change to determine what can actually be improved upon. While this certainly makes the entire process take longer than it usually does, making too many changes at once makes it difficult to determine which one of the changes led to the biggest improvement. However, if you have a lot of potential changes to test, you can then test them in groups so when something pops, you will only have to retest a specific range in order to see what caused the inspiration to strike.

Once several build, measure, and learn cycles have been completed, the product should be well on its way from moving from the initial starting point to the final, ideal phase. At some point, however, if things don't seem to be proceeding according to plan, then the question becomes whether it is better to pivot to something new or to stick with the current baseline a while longer to see what improves. The choice between the two should be relatively obvious at this point based on the data provided up to this point.

If the decision is ultimately made to pivot at this point, then it can be quite demoralizing for the team because this means going back to square one, albeit with additional data to draw on in the future. Nevertheless, issues such as vanity metrics or a flawed hypothesis can lead teams down a path that is ultimately not viable. This scenario leaves them no choice but to tear it all down and start again with an alternate hypothesis and a clean slate. It is important to try and reframe the idea of a pivot from a failure to a success because it saved the startup from potentially taking a flawed product to market and paying in a big way further down the line.

There are a few additional types of pivots as well. A Pivot that zooms in is one that takes a signal successful feature of a failed prototype and turns it into its entirely own product. A zoom out pivot, on the other hand, is when a failed prototype is useful enough to become a feature on something larger and more complicated.

The customer segment pivot occurs when the prototype proves solid, but the target audience proves to be different than anticipated. A customer need pivot occurs when it becomes clear that a more pressing problem for the customer exists, so a new product needs to be created to handle it.

A platform pivot occurs when a single application becomes so successful that it spawns an entire related ecosystem. A business architecture pivot occurs when a business switches from having low volume and high margins to high volume and low margins. A value capture pivot is one of the most extreme as it involves restructuring the entire business to generate value in a new way. The engine of growth pivot occurs when the profit structure of the startup changes to keep pace with demand.

*Small batches:* When given the option to fill a large number of envelopes with newsletters before sending them out, the common approach is to do each step in batches, fold the newsletters, place them in the envelopes, etc. However, this is actually less efficient than doing each piece by itself first, thanks to a concept known as single piece flow, a tenant of Lean manufacturing. In this instance, individual performance is not as important as the overall performance of the system. Time is said to be wasted between each step because things need to be reorganized. If the entire process is looking at a single batch, then efficiency is improved.

Yet another benefit to smaller batches is that it is easier to spot an error in the midst of them. For example, if an error was found in the way the envelopes were folded once all the newsletters had been folded, then that entire step would need to be repeated, adding even more time to the process. On the contrary, a small batch approach would determine this error the first time all the steps were completed.

*Andon cord:* The Andon Cord was used by Toyota to allow any employee on the production line to halt the entire system if a defect was discovered at any point. While this is a lot of power to give to every team member on the floor, it makes sense as the longer a defect continues through the process, the more difficult and costlier it will eventually take to remove. As such, spotting and calling attention to the problem as quickly as possible is the more efficient choice, even if it means stopping the entire production line until the issue is fixed.

*Continuous deployment:* Continuous deployment is one of the most difficult Lean Startup processes for many companies to deal with as it means constantly updating live production systems each and every day until they reach an ideal state. The essential lesson is not that everyone should be shipping fifty times per day, but that by reducing batch size you can make it through the entire build, measure, and learn cycle more quickly than your competition can. The ability to learn directly from customers is essential in this scenario as it is one of the primary competitive advantages that startups possess.

*Kanban:* This is another part of the process that is taken directly from Lean manufacturing. Kanban has four different states. The first of which is the backlog which includes the items that are ready to be worked on but have not yet been actively started on. Next is in progress, which is all of the items that were currently under development. From there, things move to build after development has finished and all the major work has been done so that it is essentially ready for the customer. Finally, the item is validated by a positive review from the customer.

A good rule of thumb is that each of the four stages, also known as buckets, should contain more than three different projects at a time. If a project has been built, for example, it cannot then move into the validation stage until there is room for it. Likewise, work cannot start on items in the backlog until the progress bucket has been cleaned out enough to free up the space. One outcome that many Lean Startups don't anticipate is that this method also makes it easier for teams to measure their productivity based on the validated learning from the customer as opposed to the number of new features being produced.

*Five whys:* Many technical issues still have a root at a human cause at some point in the process. The five whys technique makes it possible to get close to that root cause from the beginning. It is a deceptively simple plan, but one that is extremely powerful when used by the right hands. The Lean Startup system posits that most problems that are discovered tend to be the result of a lack of personal training, which on the surface can either look like a simple technical issue or even one person's mistake.

For example, with a software company, they may see a negative response from their customers regarding their most recent update. Looking more closely at the issue, it was discovered that this was due to the fact that the update accidentally disabled a popular feature. Looking closer still, this was discovered to be due to a faulty service which failed because a subsystem was used incorrectly due to an engineer that wasn't

18

trained correctly. Looking closer still, you will find that this is due to a fact that a specific manager doesn't believe in giving new engineers the full breadth of training they need because his team is overworked and everybody is needed in one capacity or another.

This type of technique can be especially useful for startups as it gives them the opportunity to determine the true optimum speed needed to make quality improvements. You could invest a huge amount in training, for example, but that doesn't mean this is always going to be the right choice at the given stage of development. However, by looking closely at the root causes of the problems in question, you can more easily determine where there are core areas that require immediate attention as opposed to solely focusing on surface issues.

Another related issue is connected to the fact that many team members are likely prone to overreacting to things at the moment, which is why the 5 Whys are useful when it comes to taking a closer look at what's really happening. There can be a tendency to use the Five Whys to point blame, at first, but the real goal of the Five Whys is to find any chronic problems caused by bad process, not bad people. This is also important to ensure that everyone is in the room together when the analysis takes place because it involves all of the people impacted by the issue, including both customer service and management. If blame has to be taken, it is important that management falls on the sword for not having a team-wide system in place to prevent the issue in the first place.

When it comes to getting started with the Five Whys, the first thing that should be focused on is instilling a feeling of trust and empowerment in the team as a whole. This means being tolerant of all mistakes the first time they happen, while at the same time making it clear that the same mistake should not happen twice. Next, it is important to focus on the system level as most mistakes are made due to a flaw in the system which means it is important to put the focus on this level when it comes to solving problems.

From there, it is important to face the truth, no matter how pleasant or unpleasant it might be. This method may bring up some unpleasantness about the company as a whole but the goal is to fix these issues, after all, and you can't fix what hasn't been brought to light. This is why it is easy to turn it into the Five Blames if you aren't careful which is why the blame should flow up in this instance. Start small and be specific. You want to get the process embedded, so start with small issues with small solutions. Focus on running the process regularly and involving as many people as you can.

Finally, it is important to designate one person on the team as the Five Whys Master. This person will be the one who is primarily in charge of seeing that change actually comes to the team. This, in turn, means they will need a fair amount of authority in order to ensure things get finished. This person will then be the one accountable for any related follow-up, determining if the system is ultimately paying off, or if it is better to cut your losses now and move on. While it can ultimately be a great way to create a more adaptive startup, it can also be harder to get into the groove of than it first appears, so it is important to look at it as a long-term investment rather than something that will be completed in the short-term.

# Chapter 2: Create a Useful Lean Startup Experiment

*Qualitative or Quantitative:* While many people assume that their startup experiment needs to be either quantitative or qualitative, the fact of the matter is that one is not inherently superior to the other. Instead, it is better to think of the two as if one was a hammer and the other was a screwdriver. While a hammer is better at putting nails in wood, that doesn't mean it is inherently superior on all fronts. Any tool can be used for good or evil, which is why it is important to focus more on validating the right metrics than it is to worry about which of these two processes is superior. In fact, using qualitative research and then validating it with quantitative research is likely going to do the most good anyway.

*Generative or Evaluative:* A generative research technique is one that doesn't start with a hypothesis per se but can still result in a wide variety of different ideas. Things like Customer Discovery Interviews fall under this type of technique. Evaluative, on the other hand, is all about testing a very specific hypothesis in order to determine a very specific result. The popular smoke test falls under this type of testing. It is perhaps this distinction, more than any other, that explains why some people end up with poor results from their experiments.

For example, a smoke test could be run to test the hypothesis that some percentage of the market will be interested in shoes that are compostable. To test this hypothesis, you would then put up a fake coming soon landing page explaining that compostable shoes are totally going to be a thing and see who signs up for the newsletter. After the work was done and the results were in, it turns out that there was about a 1 percent conversion rate when it comes to the shoes. The good news is that the hypothesis was confirmed, the bad news is that it wasn't particularly useful.

What's more, the results are unclear because it still isn't clear if the interest isn't there, if the advertising was poor, or if there is a third variable that you aren't yet aware of. This can be broadly defined as the difference between people not being interested in the value proposition and people not understanding it. The truth of the matter is that there are hundreds of reasons out there why someone might get a false negative result from a given test, just as there are a number of reasons why a false positive might be generated.

To get started, you will need to determine if the hypothesis is flawed or simply vague and, in this case, it is both. Some people are too vague when it comes to a target audience, some are a poor qualifier. As such, first, you would need to focus on a more specific demographic, and second, you would need to do research to determine how big the audience for compostable shoes would ultimately be. Only once the hypothesis is falsifiable and specific can it benefit from an evaluative experiment like the smoke test. If you can't clear up your hypothesis then you will want to start with Generative Research and work back from there.

*Market or product:* When it comes to the distinction between methods and tools, the biggest is perhaps the distinction between Product and Market. Some methods are useful when it comes to helping startups learn about their customers, their problems, and their best lines of communication. As an example, startups can listen to their potential customers to make it easier for them to understand their specific situations and what their day to day problems are like.

Other methods make it possible to learn about the product or a potential solution that will help to solve a specific problem. One good place to start is with a set of wireframes as a means of determining if the interface is as usable as it seems at face value. Unfortunately, this still won't make it clear if anyone is going to buy anything in the first place.

As these methods don't typically overlap all that well, it is important to choose one and stick with it throughout its cycle. If you combine evaluative research and generative research with Product and Market, you will end up with four different means of determining the best path forward.

Generative Market research asks questions like:

- Who is our customer?
- What are their pains?
- What job needs to be done?
- Is our customer segment too broad?
- How do we find them?

If you can't answer these questions clearly and easily, then your startup is in what is known as the Customer Discovery phase. During this phase, it is important to get to the basis of the problem prior to testing out any potential solutions to ensure that you are actually solving the right problem in the end. If you don't have a clear hypothesis to start, then you will need to generate ideas.

To do so, you may want to talk to customers to see what is bothering them or you could use a data mining approach to determine the problem, assuming you have access to enough data. You may even want to use a survey with open-ended questions if you are really fishing for ideas. Some of these methods will be qualitative and some will be quantitative, but this distinction is ultimately irrelevant in the long run. Data mining is a quantitative approach, but it helps identify problems, most famously the existence of food deserts which would have been difficult to determine in virtually any other way.

Generative Market Research Methods include:

- Surveys
- Focus groups
- Data mining
- Contextual inquiry / ethnography
- Customer Discovery Interviews

Evaluative Market experiment questions include things like:

- How much will they pay?
- How do we convince them to buy?
- How much will it cost to sell?
- Can we use scale marketing?

In order to properly evaluate a specific hypothesis, you may want to start with a landing page to determine if there is likely to be a demand. You may want to put together a basic sales pitch if you are working on a B2B enterprise type product. You could even go so far as to run a conjoint analysis as a means of further understanding the relative positioning of a few value propositions.

Evaluative market experiments that are useful if you have a clear hypothesis include:

- High bar
- Fake door
- Event
- Pocket test
- Flyers
- Pre-sales
- Sales pitch
- Landing page
- Video
- Smoke tests
- Surveys
- Data mining/market research
- Conjoint Analysis
- Comprehension – link to the tool
- 5-second tests

While this sort of research can provide lots of interesting data, it is important to keep in mind that much of it still has the potential to be wrong as signing up for a landing page is very different than actually putting money down on a product. In any situation where the customer doesn't have to commit anything more than an email address, then they don't signify an actual customer demand.

It is important to keep in mind that the value proposition and the product are not the same things. The value proposition is the benefit that your product will deliver to your target audience. As such, you cannot have a validated value proposition if you don't have a validated customer segment.

# Chapter 3: Growing a Startup

When a startup is composed of only a few people, the small team that started the company, it's easy to manage everyone and everything. You've got your first few clients, and they're happy with your work, paying all their bills on time and referring your services to other potential clients. But as your startup grows—with more staff, more clients, and more money to keep track of—it can be a challenge to manage all these aspects efficiently.

But there are ways to make this process easier so that you don't lose too much time or money. It's all about ensuring you use collaboration, effective lead generation, and strict budgeting. Here's how:

**Collaboration**

It doesn't matter if you're working with 5, 25, or 75 employees; any team, regardless of size, needs to have the right tools and resources to successfully collaborate. Teams, whether they are working alongside one another in the same space or remote, need to have awareness of the initiatives their colleagues are pursuing. Yes, there are many collaborative tools available that allow teams to message one another throughout the day and share files, but what these tools often lack is context.

Cage is a new platform that enables contextual collaboration. Through Cage, teams have the ability to gather feedback in real time, assign tasks, edit images, and distribute media files, all on one platform. By facilitating the entirety of a project, from the initial brainstorm to a final review before a video or platform is published, Cage ensures that everyone involved in the projects has full insight into updates and strategic pivots.

Regardless of the medium, every project takes on a life of its own. More often than not, facets shift over time, and these changes and discussions are often implemented across several platforms, which often lead to confusion and oversight. Cage helps teams avoid this pitfall and, as a result, empowers them to collaborate more effectively and efficiently.

## Effective lead generation

All startup founders need to focus on revenue efficiency. Otherwise, according to Forbes, you'll be totally lost. Strategic planning and tracking are the only way that you'll be able to ensure you're being as efficient as possible when it comes to revenue and effective lead generation.

LeadCrunch advises that for every dollar spent on customer acquisition, a company should generate $2.50 in return. But this is not possible if you don't prioritize top-of-the-funnel sales leads. Too many organizations waste time and money by casting large, and irrelevant nets at the top of the funnel, which results in salespeople wasting their time trying to engage customers who, simply, aren't interested. LeadCrunch's CEO, Olin Hyde, believes that the key to successful lead-generation is micro-segmenting, engagement, and nurturing.

Fueled by their AI-driven platform, LeadCrunch allows companies to pinpoint relevant, high-quality prospects, and cultivate deep relationships with them using information that is specific to their unique needs. As a result of their platform's laser-sharp focus, LeadCrunch's top performing customers often see conversion rates spanning 300-1000 percent after leveraging the platform's technology.

## Strict budgeting

It's not enough anymore for a startup to use an Excel sheet to calculate and keep track of all their budgeting needs. Even if they've moved onto Google Sheets, which are, after all, free— that's still not enough. If your startup is growing, it makes sense to invest in a B2B budgeting tool that uses actual objective data, which will help you understand how different actions and decisions affect the money coming in.

Hive9, a planning, budgeting, and analytics solution created specifically for B2B marketers, is a smart way for you to keep track of every aspect of budgeting. What's most effective about it? According to Olive & Company, "This comprehensive tool integrates your marketing goals, plan, and budget with your campaigns, so you can measure success and strategically allocate your budget...Hive9 helps you determine where your revenue is coming from, down to a specific touchpoint, and where you can improve your marketing. It also helps determine your cost per marketing lead and sales qualified lead."

According to Hive9's mission, they want to help you create a budgeting plan that you can stick to, one that directly correlates to the complex projects you have going on. They understand how stressful it can be to juggle marketing, which is ever-changing, with budget planning.

Growing your startup is certainly going to be a challenge—but it's a challenge worth taking. After all, change and growth are one of the best ways to ensure your startup thrives and succeeds. By using the strategies of collaboration, effective lead generation, and strict budgeting, you'll be able to improve your startup while also keeping control of all aspects as it grows.

Have you ever grown your startup from its original size? What challenges did you come across, and what strategies did you use to make the transition easier?

## Product development

*Listen and listen well:* Listening is fundamental to a Lean Startup. Listening is what gets you to notice your customers' needs.

It includes listening to devotees and skeptics – your fans will fuel your endeavors and reassure you of all your goodness; skeptics will provide you food for thought on angles perhaps obscured.

Listening will provide you with the "secret to unlock success" and make sense of the noise and valuable feedback that will impact your design iterations, remodeling, and execution.

Educate yourself on laws, regulations, business etiquettes, and many other considerations in your industry.

*Your first customers:* You must be a die-hard fan of your first customers. After all, they are the ones who were the early adopters, the ones who chose to trust you for the product you introduced to them. And keep selling to them again and again.

Most of the time, startups spend and are more than willing to spend to attract new customers. They forget about their existing customers and do not give importance to re-attracting, reselling, and retaining them.

As a Lean Startup company, your fastest way to feedback for

improving your product is the customer who bought it from you at the very start. They tried out your product and knew what was required to make it better for them and many others. Also, your existing happy customers are most likely to spread the word amongst their acquaintances. And referrals are the most authentic and impactful method to increase sales than many other paid programs you are likely to invest in as a startup.

Thus, get feedback from your existing customers and introduce your improved products to them to try out and give you more feedback.

Turn them into loyal customers who will set a precedent of customer value and care for your startup.

*Organic outreach:* Get to know your customers and let them know you. This is a timeless and evergreen advice for you to grow your Lean Startup faster.

Understand what an SEO audit is and invest in one for your website and social media platforms. An SEO audit will help you create and understand your startup's penetration in the market. Leverage organic outreach and give something of great value to your customers.

A simple combination of great content creation, customer feedback, and input will create a measurable impact on your organic outreach

*Content marketing:* We are avid fans of content marketing and know that content is here to stay—but only "great" content.

A great content marketing strategy and strategist will make you think of creating resources that are useful to your consumers. Great content will introduce your business properly, talk about its uniqueness, products, and how people think it is doing.

Your social media and website should have sufficient content, updated timely, and posted regularly to keep things fresh and relevant. Therefore, ensure that you create interesting content making use of the many simple tools available in the market.

A strong online presence with an effective SEO strategy and great content will not only help your startup to attract new customers but will also retain existing ones.

*Ask your customers to chip in:* A great strategy to grow fast is to involve your customers to build a prototype with you.

The lean methodology begins with a true hypothesis – a fairly accurate guesstimate to solve a problem with a product. Introduce the model to your customers and ask them to give you feedback on the usability of the product.

*Feedback:* Since Lean Startup companies are closely attached to their customers, you will relatively have more convenient time developing content and resources for your customers and prospective audience. Your business model facilitates a fluid and comfortable relationship with your market. And this close relationship can spark a valuable and insightful market research.

Ask your customers for feedback on what they think you are doing regarding selling your products, providing support, quality management, and areas for improvement. Listen carefully and make sure you create a list of their feedback and ideas.

## Customer onboarding and satisfaction

As a startup, know your customers personally. Create opportunities to meet them. Many small business brands send handwritten notes with their products to customers to make them feel welcome and part of the family. Send birthday cards, season greetings, etc. to engage your customers.

*Share stories of growth, challenges, and outcomes:* Your content marketing strategy must include the changes you make, big or small, for example, a new and improved design or a slight color variation, or updating a new phone number to manage queries to the addition of new members in your sales team, etc.

Tell your customers when you make changes to your products based on their feedback, as it will create a circle of trust and reliability which will ultimately affect your bottom line.

Sharing information this way will establish a culture code in your startup and communicate your value proposition via meaningful actions to both your customers and prospective customers.

*Accept mistakes:* There will be mistakes made on the way. Own them and seek to clarify and apologize appropriately.

## Paid outreach

At times, paid marketing and outreach can work brilliantly for a startup. However, you will need to think through before executing this strategy.

The sure sign of a well-paid outreach is profit. You should be making more money with paid outreach per customer than you would in acquiring one. Even if the ratio is 2:1, paid outreach is working.

By the end of the day, your goal is to make customers happy. When your customers are happy with your product, they will refer you to their friends and family, and you will continue to succeed in your endeavors.

And this leads us to our suggestion of looking into influencer marketing. Influencer marketing will involve your happy customers to become your brand ambassadors and talk about your product in their social circles to co-build your brand.

It is a win-win for you, your existing customers, and your potential customers who will have been influenced by their friends or influencers. This way, your startup's value proposition travels smoothly into the market.

## Build a great team

Keep good people around you. Keep people who know what they are doing. Keep people who thrive on feedback and know how to use that feedback for growth and improvement; this is key to the interactive model of improving your product and keeping it relevant to your customers' needs.

*Learn from examples:* These are some examples of well-known companies who did not want to waste their time, their customers' time, or that of their investors. They got down to developing the product that would fulfill the needs of their customers.

When Dropbox came across the lean methodology, the company began thinking about their product which at the time of its inception was competing with many other similar products.

They learned how to test new products with their customers and incorporated their feedback. From 100,000 registered users, Dropbox went over 4,000,000 in only 15 months with the Lean Startup principles.

The giant General Eclectic also went into the Lean methodology to develop faster solutions with FastWorks - a complete mindset changing program to how things usually worked at GE. A very popular example was the development of a gas turbine two years faster with 40% less cost. The lean methodology is not only for startups after all!

After approaching local shoe stores and grabbing photos of their shoes with consent, the owner of Zappos tested his hypothesis of whether an online shoe-selling site would work. And it did! Zappos is the largest online shoe store with over 1000 brands featured.

The ideas above are fairly consistent with the ideas you hear every day. You learned these at business schools, through books and blogs, and friends and colleagues. Simple as they may sound, it is always the simpler ideas that are key to great outcomes.

# Chapter 4: Six Sigma Basics

Six Sigma is the name of a lean system for measuring quality with the ultimate goal of getting as close to perfection as feasibly possible. Ideally, a company that is running at Six Sigma maximum efficiency would generate, at most only 3.4 defects per million attempts at a particular process. Zshift is the name given to these deviations which show the difference between a poorly completed process and a perfectly completed one. The baseline Zshift is 4.5 while the ideal value is 6. Processes that have not yet been analyzed via the Six Sigma process typically score somewhere between 1 and 2.

*Levels of Zshift:* If the Six Sigma analysis of a process is at 1, then this means you can expect customers to get exactly what they want somewhere around 30 percent of the time. If this is increased to a Zshift 2, then you can expect the process to give the customers exactly what they are looking for about 70 percent of the time. If the process reaches a 3, then it will be accurate about 93 percent of the time, a 4 will be 99 percent accurate, and a 5 and 6 reach even smaller divisions towards the goal of 100 percent accuracy and customer satisfaction.

*Six Sigma cert:* Within Six Sigma, there are a variety of different certification levels that can be achieved, each with its own tasks and responsibilities as it relates to the whole. Six Sigma is all about decreasing the risk of production errors by reducing waste and improving efficiency. The first two levels of certification, White and Yellow Belts, are crucial to this part of the process as, while working under higher level Sigmas, they do things like ensure the data that is coming in is on the right track and carry out specific functions that are designed to add overall value to the process. These certificates are also a great way to be exposed to the overall Six Sigma methodology.

The next level of certification is the Green Belt which allows holders to work more directly on Six Sigma projects being helped by those above them while also allowing them to oversee projects being handled by Yellow and White Belts. Black Belts then lead high-level projects while also supporting and monitoring those at other tiers. Finally, Master Black Belts are those who are often brought in specifically to start Six Sigma at a company and are knowledgeable to mentor everyone at every level.

**Six Sigma is the tool for you if...**

While almost any company and any team can benefit from the Six Sigma in some shape or form, this doesn't mean it is always going to be the right choice, especially for a startup company. In fact, deciding if it is the right choice or not actually depend on a wide variety of different specifics, including how committed the team is to implementing the system in the first place and what the company's developing culture is like.

*Leadership involvement is key:* When discussing the idea of transitioning to Six Sigma, it is important to not look at it as another type of flash in the pan management style that is going to go out of fashion as quickly as it appeared. Instead, it is far more likely to find purchase if it is pitched as an enhancement of an existing system. Likewise, you can always find opposition to something new from someone in the company which is why it is important to start with buy-in from the top and work your way down the list from there. It is extremely important to have the full management team on board from the start as if it doesn't look like there is a consensus regarding the new system, then it is going to be dead in the water before it even gets off the ground.

This doesn't mean that the entire team needs to be committed to the idea of Six Sigma from the start, but it does mean that it is vital that the change that is on the horizon that needs to be institutional which means the leadership needs to put forth a united front. Don't forget, the human brain is a creature of habit which means that it will recoil from new systems that seem too complicated if the system is, at all, perceived as optional. As such, if there is an opposition to the new system, stress the idea that it is important it be expressed in private.

*Consider the infrastructure:* Six Sigma is all about leaders mentoring those on their teams in order to make Six Sigma work as effectively as possible. As such, if you hope to transition to a Six Sigma system successfully, then this needs to be a full-time job for at least one person, at least until the new, positive, habits have formed for good. While this might not seem cost-effective in the short-term the Six Sigma savings that will appear once the system is up and running properly are sure to more than make up for it.

*Consider what will boost compliance:* After support of management has been assured and the infrastructure is in place to make the project really pop, the next step will be to ensure that the rest of the team members have a motivational reason to fall in line. While active rewards aren't going to be required once the Six Sigma process has been properly internalized, they are a good way to help the team get used to looking at problems as though Six Sigma is the solution. While the right way to track team progress is going to be different for every team, it is vital that each member of the team feels an immediate and compelling reason to commit to the new program, at least at first.

After all, companies are like any other body that is currently in motion, the bigger the company, the more inertia is needed when making large changes. This is where your startup has the opportunity to outmaneuver the big boys and start gaining some ground as quickly as possible.

*Who else is using Six Sigma in your field:* While Six Sigma has proven its worth in a wide variety of different fields, this doesn't mean that each of these fields is going to be ready to adopt the process with open arms immediately. While looking to the future is one thing, if your startup is also going to the first company to adopt Six Sigma as a common practice in its industry, then you need to be prepared from extreme resistance from every side and especially any members of the old guard that you may have brought on to ensure the real work gets done. Luckily, the science behind Six Sigma is solid which is why it should be fairly easy to come up with specific examples of how it can help your company to silence any opposition.

*Consider training objectives:* Depending on the size of your team, training everyone as a whole might make sense. Eventually, however, different training levels are going to need to be enforced, as not everyone is going to need to be a Black Belt. As such, it is important to look more closely at the various levels and the qualifications for each before determining how training can best be broken up for maximum effectiveness. It is important to also factor in how the training will affect any other duties the trainees might have as well as what areas are going to be focused on most stringently.

Taking the time to work out the specifics of your training scenario before you get started is sure to make all the difference in the world when it comes to implementing Six Sigma successfully. Remember, there are no one-size-fits-all options in this scenario which means planning out the specifics of your team's training could literally be the success and failure of the entire project. What's more, it can also make other issues more apparent when otherwise, they would not have been noticed until training was already underway.

*Consider flagship projects:* After the Six Sigma training is out of the way, it is important to have a few important projects waiting in the wings in order to show the team that the system is worth it. Not only will these projects help to get the entire team excited about Six Sigma, but they will also be useful down the line if questions as to whether or not Six Sigma is worth keeping start popping up as well. As a general rule, you will want to start with at least a Green or Black belt project and then do everything in your power in order to ensure they end up being successful.

While doing so, it is also important to not spread the Black Belts and Green Belts that you have on your team too thin that it makes them struggle to match their deliverables. Instead, it is better to have too many people on a few projects to ensure they are completed to perfection. Remember, if things go according to plan then you will have plenty of time to complete other projects once these go over like gangbusters.

Furthermore, it is important that these early chapters are more than just fluff projects, they need to be things that are legitimately beneficial to the company as a whole. If your early projects are heavily publicized but do little to produce viable results, then you run the risk of Six Sigma being seen as little more than a fad with lots of flash and little substance. You can ensure that this doesn't happen by instead choosing projects that have a clear value, regardless of whether the person making the decisions is trained in Six Sigma or not. Remember, public opinion is one of the most important resources to covet at this stage and will continue to be so until Six Sigma has become a habit for the entire team.

# Chapter 5: Implementing Six Sigma

*Give the team a reason to want to try Six Sigma:* In order to ensure that Six Sigma is implemented successfully, it is vital that you take the time to motivate your team in the most effective way possible, so they understand why it is so important to adopt the Six Sigma methodology. Depending on the state your startup company is in, the burning platform scenario might be the best choice.

The burning platform is a type of motivational technique whereby you explain that the situation your company finds itself in is just as perilous as standing on a burning platform and the only way to turn things around for good is by implementing Six Sigma. It is important to have stats that back this idea up, though fudging the numbers for productivities sake might not hurt either. Adapting to Six Sigma can be difficult for team members who are set in their ways and external motivation may be just what the doctor ordered.

*Give team members the tools they need:* After the primary round of Six Sigma training has finished, it will be important to ensure that you have a strong mentorship program in place, along with details on the finer points of the process for those who need them. One of the worst things that can happen at this point is for a team member to express an interest in the program only to become disinterested when additional materials are not readily available. A team member who cannot easily find answers to their questions is a team member who will not follow Six Sigma processes when it really counts.

*Prioritize properly:* Regardless of the situation, there are always going to be a variety of potential outcomes. While talking and planning for Six Sigma is one thing, taking steps to actively prioritize it is another entirely. When team members see those in leadership roles prioritize Six Sigma outcomes, it makes them more likely to prioritize Six Sigma activities in their own jobs as well. Additionally, it is important to make it clear that quality is critical, as is listening to the customer when it comes to ensuring Six Sigma leads to measurable results so that team members of all level of certification can keep an eye on the overall progress the company is making.

*Make it a group thing:* When it comes to tutoring your team about Six Sigma, it is important to ensure that they make personal connections with how it will affect their jobs for the better so that they feel more personally invested in the program's overall success. This may come about by ensuring the entire team is able to provide buy-in or making different team members responsible for enforcing different aspects of the Six Sigma process as taking the time to ensure that everyone feels connected to seeing Six Sigma succeed will ensure personal retention rates remain as high as possible.

*Track the results:* Determining a realistic metric that can determine appropriate levels of success before and after Six Sigma is an important step in the process as it can provide you with the motivating data that is required to ensure that Six Sigma adoption is at an all-time high. On the other hand, if it turns out that the system actually ends up being ineffective, then you will be the first to know as well. Regardless, having a viable metric to properly determine aptitude is sure to come in handy more than once. What's more, assuming the results are positive then it is sure to be a great motivating factor for the entire team and provide yet another reason why sticking with Six Sigma is so important.

*Reward team players:* While offering viable reasons for the team to adopt Six Sigma is one thing, it is still important to provide positive reinforcement during the early days so that everyone is constantly motivated to follow the Six Sigma process until it becomes a habit. The goal here should be to choose a reward that is valuable while at the same time not being so extravagant that eventually removing it won't completely remove the team's desire to keep up the good work.

## Six Sigma criticism

*Six Sigma is just a fad:* While it has only been back in the spotlight for a few years, the fact of the matter is that the origins of Six Sigma can be traced all the way back to the early 1900s when it was used by entrepreneurs like Henry Ford, Edward Deming, and Walter Shewhart. Additionally, it further separates itself from true fad management styles by being more focused on the use of data as a means of ensuring the best decision is made at the moment as possible, specifically those with a focus on the customer as a means of ensuring a viable return on every investment.

*Switching to Six Sigma is resource intensive:* While it's true that training the team in the Six Sigma process is time-consuming, the end goal is for it to save far more time than the training will cost in the long run by ensuring team members do their jobs as effectively as possible moving forward.

It is important to remember the story of the pair of lumberjacks who worked day after day in the forest. One man worked himself to the point of exhaustion every day while the other man spent the time preparing properly, and at the end of the day, both men had always chopped the same amount of wood. If your team has the opportunity to work smarter instead of harder, why wouldn't you provide them with the tools they need to make that the new norm.

Furthermore, the cost of training the team the Six Sigma process can be further mitigated over time by spreading out the training courses as required. While this means the team won't start seeing the results as quickly as might otherwise be the case, even getting the entire team up to Yellow Belt will produce noticeable results. Furthermore, any funds put towards this type of training can really be seen as an investment in the business as a whole and should be treated accordingly.

*Our team is too small for Six Sigma to be effective:* While the effectiveness of Six Sigma is proportionate to the inefficiency of the processes previously in place, that doesn't mean it doesn't have something to offer companies that are just getting up and running as well. After all, Six Sigma offers a different way of looking at the types of business interactions that happen day to day in hopes of increasing productivity and, as a result, profits as well, regardless of the size of the team that is utilizing it. What's more, smaller teams will actually be able to take on the Six Sigma mantle more easily than larger companies as the number of resources required to train 10 team members are always going to be far lower than what it would cost to train 50 instead.

Furthermore, smaller businesses can be hindered more by production bottlenecks which means a Six Sigma system would potentially be able to lead to greater periods of growth as issues that may not otherwise have been addressed for years are taken care of before they become institutional problems. Regardless of the size of the company in question, taking the time to truly streamline relevant processes or improve customer relations is always going to be the right choice.

*Six Sigma doesn't apply here:* While it's true that Six Sigma isn't going to apply to every single industry across the board, it has moved beyond its manufacturing sector roots. Furthermore, studies show that industries that provide services are prone to more waste than the manufacturing sector in the first place. This is due to the fact that so much of what is produced is intangible that it makes standardizing any process extremely difficult. This is where Six Sigma comes in as it has plenty of processes in place to track the services that are being provided which can ultimately be used to improve efficiency.

*Six Sigma is difficult to use practically:* While it may have a reputation for being all about the numbers, a vast majority of the tools and principles that are used in implementing Six Sigma require less math and more common sense. As an example, consider the mitigation of waste which is one of the most important aspects of Six Sigma, and something that only requires an understanding of the business in question and how it can be done in a more effective fashion. This is indicative of most of Six Sigma which is largely about fostering the mentality that makes it possible for team members to find the root cause of an issue, regardless of how long it might take. While formulas and mathematical equations may be used, they are simply a justification for this fact.

*Lean is plenty for now:* When expressing the benefits of Six Sigma, it is important to make it clear that it is a variation of the Lean system, not a replacement for it. In fact, the system is often referred to as Lean Six Sigma. When used in conjunction with one another, Lean will then the throughput and speed of the process and simplifying to ensure that the team is able to do the best with what they have available. Six Sigma then takes these improved processes and makes them of the highest quality possible by reducing defects and, as a result, lowers the deviation. Combining the two can only lead to better results overall.

# Chapter 6: Additional Strategies

## Kaizen

The word Kaizen translates to "continuous improvement" which is obviously an important goal when it comes to creating the most effective Lean system possible. The goal of the Kaizen strategy is to ensure that all of the talent within the team is always focusing on improving whenever and wherever possible. This strategy is relatively unique for a Lean strategy in that it is more than just a direct plan of action, it is also a general philosophy for the company as a whole. The goal of Kaizen is to create a culture that is supportive of improvement in all of its forms while also creating groups that are focused directly on improving key processes and reaching well-defined goals.

Kaizen is a great strategy to implement while you are standardizing your work process as the two complement one another well. Standard practices lead to current best practices which Kaizen can then improve upon. The Kaizen strategy can be useful when it comes to improving any strategy that your team uses with any real degree of regularity as long as you are fully aware of the end goal for the updated process. From there, you will need to review the current state of things before adding any improvements. From there it is just a matter of following up properly in order to ensure any proffered improvements work as expected.

Training the team in Kaizen actually serves double duty as it teaches them to apply the philosophy and the plan of action at the same time. This type of thinking is often habitually formed by those who are constantly looking for ways to improve their most commonly used processes while also allowing other team members to approach common tasks in new and innovative ways as required. This mindset should naturally be nurtured whenever possible as it is the only way to ensure more fruitful results in the long run.

While constantly improving existing practices is a great place to start, it is important that the Kaizen your team is practicing does not only occur after the fact. When new processes are created, it is best for everyone that they are held to the same examination process as any other. Hindsight is useful, foresight gets results.

*Kaizen steps*

- The first thing you will need to do is to standardize your process, not just the process that you are looking to put through the Kaizen process but all the processes to ensure that any eventual improvements are as beneficial as possible.

- Next, you will need to compare the processes at play in order to determine where steps that are being used in some processes can be used successfully elsewhere as well. When taking this step, it is vital that you look at true KPIs as opposed to anecdotal information for this step as, otherwise, it can be easy to get off on wrong track without even realizing it.

- After you have determined where real change should occur, the next step is to consider what you currently have available to make completing the process as easy as possible. During this period, you will want to consider the start of the project as well as its conclusion and then brainstorm all the possible ways to reach point B from point A. While no idea should be off the table at first, it is important that you ensure you only move forward with ideas that are truly useful as well as innovative as innovation for innovation's sake is only going to create waste.

- The final step is going to be to repeat as needed so that new innovations become standard operating procedures so that you can then begin the entire process anew. When it comes to Kaizen, the only bad idea is to rest on your laurels.

*Creating a Kaizen mindset:* Getting the entire team together for a Kaizen event where everyone brainstorms ways to streamline a specific process is relatively straightforward. However, training your team to always work from a Kaizen mindset can be a far more difficult task. Difficult does not mean impossible, however, and the best way to start to train them to this improved way of thinking is to focus on creating a corporate culture where elimination of waste is everyone's top priority. If you can keep this idea in the team's mindset, during every meeting, every performance review, every informal conversation, day in and day out, then eventually team members will start noticing waste without even having to consciously think about it. Once this occurs they will be well on their way to finding ways to work around it instead.

With this done, you will also want to start to set aside a specific time each day to allow team members to look at the processes

49

they use every single day and do nothing else but really think hard about them. It is important to always remember that the human mind loves repetition almost as much as it loves patterns which is why it is so easy to follow the steps for a process you have done a hundred times without even thinking about it. While this can make the process go faster if the steps involved are optimized, it can also make it easy to complete the process with blinders on and not notice points of inefficiency while you are in the midst of them. As such, providing the team with the time they need to think about their processes separate from actually doing them will let them look at the entire project from a different angle.

If you take this exercise a step further, you will then provide the team with an opportunity to talk to the rest of the team about their processes as well. This cross-pollination of ideas will give each process an entirely fresh set of eyes which will provide insight into even more blind spots. This is especially useful for particularly complex processes, just make sure that everyone takes detailed notes, so nothing gets lost in the shuffle. Additionally, it is important to emphasize that there are no wrong answers during this stage, a free and open dialogue can provide solutions to problems that you previously weren't even aware you were facing.

**Poka-Yoke**

The Lean system strategy known as Poka-Yoke is most accurately translated as actively guarding against mistakes. Essentially, Poka-Yoke can be thought of as a variety of failsafe procedures that are naturally built into any processes specifically for the purpose of catching common errors. Poka-yoke is best used on tasks that are especially repetitive, require precise repetition across numerous steps, or require an extreme period of focus to use correctly. This tool is an especially beneficial type of Muda that works to ensure the overall value while not necessarily creating any of its own.

When Poka-Yoke is at its most effective, it relies on a thorough understanding of the steps in every process as well as additional ways of mitigating potential pain points as effectively and cheaply as possible, while also taking care not to create any new bottlenecks as a result.

*Control Poka-Yoke:* Control Poka-Yoke does not allow the next step of the process to move forward until a found error has been corrected. As an example, the way a USB dongle is designed so that you cannot plug it in unless it is facing the right way is a Control Poka-Yoke as it ensures you cannot plug in the device in such a way that it will not work once you do so.

*Warning Poka-Yoke:* Warning Poka-Yoke, as the name implies, provides the team member completing the process that they made an error on a proceeding step.

*Contact method Poke-Yoke:* The contact method Poke-Yoke works under an assumption that a third party, either a device or a person, is monitoring the steps that are being taken to ensure no errors materialize. A spell-check program is a good example of this type of Poka-Yoke. Contact method Poka-Yoke is especially useful if the same task needs to be repeated as quickly as possible in order for the process to run smoothly.

In order to determine where Poka-Yoke can be of the most use to your processes, the first thing you will need to do is to determine which steps in the process already cause the most harm, or have the most potential to cause harm, in the shortest period of time overall. You may want to start by determining the processes' critical features and then looking at potential causes of failure before determining a signal method that will work effectively in the situation.

*Fixed Value Poka-Yoke:* This type of Poka-Yoke is useful in situations where the overall process is quite short but requires the same step in the process to be run numerous times in a row. Poka-Yoke is useful in scenarios like this as they allow the person who is completing the process to know how many times they have repeated a specific step. As an example, think of an administrative assistant making numerous copies of a document who first counts out the amount of paper they need to ensure they don't have to count each as it is made.

*Motion step Poka-Yoke:* This type of Poka-Yoke is useful in situations where a team member needs to perform numerous different tasks, in a specific order, many times. This Poka-Yoke determines when specific steps have been completed to ensure the team member completing the process remains on track. As an example, consider any website where you are asked to enter your payment information. When the website tells you that you haven't entered the correct payment details and can also track that you haven't yet checked the box to prove you aren't a robot, then it is an example of motion step Poka-Yoke.

*Self-Check Poka-Yoke:* This type of Poka-Yoke requires the fewest additional resources to complete and instead just requires a little extra time to give the team member performing the process the opportunity to check their work before they move on to each new step. This is a good choice for scenarios where mistakes are extremely obvious, and its biggest drawback is that it requires extra time during which the team member must remain focused as well.

*Task Poka-Yoke:* This type of Poka-Yoke is useful when it comes to processes that require a team member to directly come into contact with a customer as it helps cut down on mistakes that are made in live situations. A great example of this is the change machines at grocery stores that prevent cashiers from making mistakes when counting out change.

*Treatment Poka-Yoke:* This type of Poka-Yoke works to ensure that the customer always has a positive and efficient interaction whenever they encounter a team member while working through the course of a specific process. The goal here is to standardize what team members say as much as possible in an effort to prevent any potential mistakes before they happen. This is especially useful for new businesses as it gives new team members something to fall back on when they encounter something new, which is basically everything at this point. A great example of this type of Poka-Yoke is the scripts call centers use.

*Tangible Poka-Yoke:* This type of Poka-Yoke aims to standardize the physical element of the customer's experience. In situations where individual customers have widely varying needs, this type of Poka-Yoke is often the best way to standardize service. A good example of this type of Poka-Yoke is uniforms.

*Preparation Poka-Yoke:* This type of Poka-Yoke: aims to work with the customer directly to influence expectations and goals prior to the experience. Depending on the requirements surrounding the products or service in your business, this can be a great way to make sure that customers are prepared properly before they speak to a team member to ensure the whole process goes as smoothly as possible. A good example of this type of Poka-Yoke are menus that are visible to patrons of fast food restaurants before they order so the actual order process proceeds as smoothly as possible.

# Conclusion

Thank you for making it through to the end of *Lean Startup: The Complete Step-by-Step Lean Six Sigma Startup Guide*, let's hope it was informative and able to provide you with all of the tools you need to achieve your goals. Just because you've finished this book doesn't mean there is nothing left to learn on the topic; expanding your horizons is the only way to find the mastery you seek. Additionally, it is important to keep in mind that, while there is some overlap between any two startups, much of what is going to take place is going to be largely unique to the startup in question.

After all, isn't the point of a startup to do something new? As such, it is important to understand that while following the Lean Startup strategy can certainly lead to success, sometimes you may have to make your own way because what you are trying to do is so out there that the existing methods of testing don't apply. This doesn't mean that you should abandon all that the Lean way of doing things has done for you thus far, it simply means that you will need to take what you have learned and use that to create logical ways to test whatever it is you are prototyping. Likewise, it is important to not get impatient and try to rush the process. After all, creating a viable product or service that a targeted portion of the audience is interested in is a marathon and not a sprint which means the slow and steady wins the race.

Finally, if you found this book useful in any way, a review on Amazon is always appreciated!

# Lean Analytics

*The Complete Guide to Using Data to Track, Optimize and Build a Better and Faster Startup Business*

# Introduction

Congratulations on getting a copy of *Lean Analytics: The Complete Guide to Using Data to Track, Optimize and Build a Better and Faster Startup Business* and thank you for doing so.

The following chapters will cover everything you need to know to get started with the process of Lean Analytics. The Lean Support system is a great way to ensure that your business is as efficient as possible by eliminating the amount of waste that is present. The Lean Analytics section is going to help with data collection and analysis. Thus, you'll determine where the waste is present, and this will help you to pick the right metrics to implement.

This book will discuss Lean Analytics and how its processes can help you reduce waste and find the best strategy to improve your business. It is just one step in the Lean Support System, but it is an extremely critical step. This guidebook will provide you with the information that you need to get started so that you can become an expert in Lean Analytics in no time.

There are plenty of books on this subject on the market, so thanks again for choosing this one! Please enjoy!

# Chapter 1: What is Lean Analytics?

The central idea behind Lean Analytics is on enabling a business to track and then optimize the metric that will matter the most to their initiative, project, or current product.

There is often a myriad of methods to improve your product, but you may not have the time to work on all of them. With Lean Analytics, you will learn how to find and address the one thing that will make the biggest difference.

Setting the goal of focusing on the right method will help you see real results. Just because your business has the ability and the tools to track many things at once, does not mean that it would be in your best interest to do so.

Tracking several types of data simultaneously can be a great waste of energy and resources and may distract you from the actual problems. Instead, you will want to focus your energy on determining that one vital metric. This metric will make the difference in the product or service that you provide.

The method in your search for this metric will vary depending on your field of business and several other factors. The way that you'll find this metric is through an in-depth understanding of two factors:

- The business or the project on which you're presently working.
- The stage of innovation that you are currently in.

Now that we have a basic understanding of Lean Analytics and what it means let's take some time to further explore and see its different parts.

### What is Lean?

Lean is a method that is used to help improve a process or a product on a continuous basis. This works to eliminate the waste of energy and resources in all your endeavors. It is based on the idea of constant respect for people and your customers, as well as the goal of continuously working on incremental improvements to better your business.

Lean is a methodology that is vast and covers many aspects of business. This guidebook will spend sufficient time discussing a specific part of Lean, Lean Analytics. Here, you can learn how to make the right changes. Of course, you will need a working understanding of where to start, and Lean Analytics can help.

Lean is a method that was originally implemented for manufacturing. The idea was to try to eliminate wastes of all kinds in a business, allowing them to provide great customer service and a great product while increasing profits at the same time. Despite its beginnings, the Lean methodology has expanded to work in almost any kind of business. As long as you provide a product or a service to a customer, you can use the Lean methodology to help improve efficiency and profits.

For instance, how will you determine which metric will help you succeed? Which metric will prove to be the best and result in the most improvement compared to others? How will the metric help, how should it be implemented, and how can you ascertain if it's successful in the end? Lean Analytics can help you gather the necessary information to find and work with the right metric.

## Lean Analytics

Lean Analytics is part of the methodology for a lean startup, and it consists of three elements: building, measuring, and learning. These elements are going to form up a Lean Analytics Cycle of product development, which will quickly build up to an MVP, or Minimum Viable Product. When done properly, it can help you to make smart decisions provided you use the measurements that are accurate with Lean Analytics.

Remember, Lean Analytics is just a part of the Lean startup methodology. Thus, it will only cover a part of the entire Lean methodology. Specifically, Lean Analytics will focus on the part of the cycle that discusses measurements and learning.

It is never a good idea to just jump in and hope that things turn out well for you. The Lean methodology is all about experimenting and finding out exactly what your customers want. This helps you to feel confident that you are providing your customers with a product you know they want. Lean Analytics is an important step to ensuring that you get all the information you need to make these important decisions.

Before your company decides to apply this methodology, you must clearly know what you need to track, why you are tracking it, and the techniques you are using to track it.

## Focus on the fundamentals

There are several principles of Lean that you will need to focus on when you work with Lean Analytics. These include:

- A strive for perfection
- A system for pull through
- Maintain the flow of the business
- Work to improve the value stream by purging all types of waste
- Respect and engage the people or the customers
- Focus on delivering as much value to the customer as effectively as possible

## *Waste and the Lean System*

One of the most significant things that you will be addressing with Lean Analytics, or with any of the other parts of the Lean methodology, is waste. Waste is going to cost a company time and money and often frustrates the customer in the process. Whether it is because of product construction, defects, overproduction, or poor customer service, it ends up harming the company's bottom line.

There are several different types of waste that you will address when working with the Lean system. The most common types that you will encounter with your Lean Analytics include:

- **Logistics:** Take a look at the way the business handles the transportation of the service or product. You can see if there are is unnecessary movement of information, materials, or parts in the different sections of the process. These unnecessary steps and movements can end up costing your business a lot of money, especially if they are repeated on a regular basis. This will help you see if more efficient methods exist.

- **Waiting:** Are facilities, systems, parts, or people idle? Do people spend much of their time without tasks despite the availability of work or do facilities stay empty? Inefficient conditions can cost the business a lot of money while each part waits for the work cycle to finish. You want to make sure that your workers are taking the optimal steps to get the work done, without having to waste time and energy.

- **Overproduction:** Here, you'll need to take a look at customer demand and determine whether production matches this demand or is in excess. Check if the creation of the product is faster or in a larger quantity than the customer's demand. Any time that you make more products than the customer needs, you are going to run into trouble with spending too much on those products. As a business, you need to learn what your customer wants and needs, so you make just the amount that you can sell.

- **Defects:** Determine the parts of the process that may result in an unacceptable product or service for the customer. If defects do exist, decide whether you should refocus to ensure that money is not lost.

- **Inventory:** Take a look at the entire inventory, including both finished and unfinished products. Check for any pending work, raw materials, or finished goods that are not being used and do not have value to them.

- **Movement:** You can also look to see if there is any wasted movement, particularly with goods, equipment, people, and materials. If there is, can you find ways to reduce this waste to help save money?

- **Extra processing:** Look into any existing extra work, and how much is performed beyond the standard that is required by the customer. Extra processing can ensure that you are not putting in any more time and money than what is needed.

### How Lean can help you define and then improve a value stream

Any time that you look at the value stream, you will see all the information, people, materials, and activities that need to flow and cooperate to provide value to your customers. You need these to come together well so that the customer gets the value they expect, and at the time and way, they want it. Identifying the value stream will be possible by using a value stream map.

You can improve your value stream with the Plan-Do-Check-Act process. This strategy can be used upfront so that you can design the right processes and products before they reach their finished form. Additionally, the strategy helps you to create an environment that is safe and orderly and allows easy detection of any waste.

Another method of creating this environment is the 5S+ (Five S plus): sort, straighten, scrub, systematize, and standardize. Afterward, ensure that any unsafe conditions along the way are eliminated.

The reason that you will want to do the sorting and cleaning is to make it easier to detect any waste. When everything is a mess, and everyone is having trouble figuring out what goes where, sorting and cleaning can address waste quite fast. There will also be times when you deem something as waste and then find out that it is actually important.

When everything is straightened out, you can make more sense of the processes in front of you. Afterward, you can take some time to look deeper into the system and eliminate anything that might be considered as waste or unsafe, and spend your time and money on parts of the process that actually provide value for your customer.

# Chapter 2: The Lean Analytic Stages Each Company Needs to Follow

To be successful with Lean Analytics, you'll need to follow several different stages. You won't be able to move on to the next stage if you do not complete the preceding step. There are five in particular that you will need to focus on to get work done with this section of the Lean support methodology. The five stages are:

- **Stage 1:** The initial stage is where you will concentrate on finding the problem for which people are searching for a solution. A business that focuses on business to business selling is going to find this stage critical. When you address this problem, then you can move on to the next stage.

- **Stage 2:** For this stage, you are going to create an MVP product that can be used by early adopter customers. This stage is where you are aiming for user retention and engagement, and you can spend some time learning how this will happen when people start to use the product. You can also learn this information based on how the customer uses your site and how long they stay. You'll take some time at this stage because you will need to experiment and also may need to go through and choose from a few different products before you get the one that is right for you. Once you have this information, you can move on.

- **Stage 3:** Once you find out how the early adopter customers are going to respond to a product or service, it is time to find the most cost-efficient way to reach more customers. Once you have a plan ready to get those customers, and then more of them start purchasing the product, then you can move to the next stage. You would not want to go with a product that may be popular but costs a ton of money, which will cut into your revenues and can make it difficult to keep growing in the future.

- **Stage 4:** You are now going to spend some time on economics and focusing on how much revenue you are making. You want to be able to optimize the revenue, so you need to calculate out the LTV:CAC ratio. LTV is the revenue that you expect to get from the customer, and the CAC is the cost that you incurred to acquire that customer. You can find this ratio by dividing your LTV by the CAC. Your margins are doing well if you get an LTV that is three times higher than the CAC. The higher the margins you get, the better because that means you are going to earn more in profits from the endeavor.

- **Stage 5:** In the final stage, you will then take actions that are necessary to grow the business. You can continue with your current plan if you are making a high enough margin from the previous steps, or you may need to make some changes to ensure that you can earn enough revenue to keep the business growing. You can also spend time making plans on where you would like to concentrate on in the future to increase the growth of your business and help it expand. The main goal for your business is to keep growing and increase revenue. This step helps you to reevaluate what you have in your current plan and decide if it is working for you or if you need to go with a different option.

# Chapter 3: The Lean Analytics Cycle

The Lean Analytics Cycle is vital in helping you get started on this part of the Lean support methodology with your business. There are four steps that will come with this process, and following each one can be crucial in ensuring that this works for you.

The best way to think about the Lean Analytics Cycle is like the scientific method. You need to do some thinking to determine what needs to be improved in your business, form a hypothesis to help lead your findings, and then perform experiments to see if that is the right process for you to keep following. If things don't work out, you don't just give up. You will continue to find new experiments, going with the same hypothesis if it works (otherwise you'll need to form a new hypothesis) until you find the right solution.

The Lean Analytics Cycle will be incredibly helpful when you begin going through the entire process. Let's take a look at the steps that you need to fulfill to use the Lean Analytics Cycle.

### *What do I need to improve?*

Before you can do anything with the Lean Analytics Cycle, you must really understand your business. You need to know all the important aspects of your business, in addition to knowing what you want to change.

During this first step, you may need to talk to other businessmen to help you find what metric you should use, based on what is most relevant to your business right now. You may also want to take a look at your business model to find out what metric will work best for you.

After you have time to choose a metric, you should connect it to the KPI or the Key Performance Indicator. An example of this is the metric that is seen as a conversion rate if the KPI is the number of people who currently purchase the product.

To make this step easier, the first thing that you would want to do is write down three metrics that are important for your business. Afterward, write down the KPI that would be measured for each metric.

Never try to implement the Lean system without understanding the most important processes that need to be improved. Sure, you could probably make a long list of things that you may want to improve in your business. But you won't really see the benefits of the Lean system if you don't pick things that are important to the overall functioning of your business. Look closely at what your business needs to improve, and pick the one that is the most important before moving on.

### *Form a hypothesis*

This is a stage where a level of creativity needs to come into play. The hypothesis is going to give you the answers that you need to move forward. You will need to look for inspiration, and you can find it in one of two ways. You can look for an answer for something like "If I perform ____, I believe _____ will happen, and _____ will be the outcome."

The first place you can look into is any data that you have available. Often, this data will provide you with the answer that you need. If you do not have data at all, you may need to do some studying of your own to come up with an answer. You could use some of the strategies from your competitors, follow the practices that have worked well for others, do a survey, or study the market to see what the best option will be.

What you need to keep in mind here is that the hypothesis is there to help you to think like your audience. You want to keep asking questions until you understand what they are thinking, or learn to understand the behavior of your audience or customer.

### *Conduct an experiment*

After you have taken the time to form a hypothesis, it is time to test it out with the help of an experiment. There are three questions that you need to consider to get started with an experiment:

- **Who is the target audience?** You need to carefully consider who your customers are and whether or not they are the right customers, or if you should look somewhere else to get better results. Also, think about some of the ways that you reach them, and if there are better ways to do this.

- **What do you expect the target audience to do?** This often includes purchasing the product, using the product, or something similar. You can then figure out if the audience understands what you want them to do; is it easy for them to do this action, and how many of the target audience are completing the task?

- **Why do you think they should accomplish the action**? Are you providing them with the right motivation to accomplish the task? Do you think that the strategy is working? If they aren't being motivated enough by you, are they doing these things for the competitors or otherwise?

Answering these questions is vital because they may help you understand your customer better than ever before. Creating your experiment during this stage does not have to be difficult. Try using the following sentence to help you get started:

"WHO will do WHAT because WHY to improve your KPI towards the defined goals or target."

If you have gone through and come up with a good hypothesis in the previous step, then it shouldn't be too hard to create a good experiment as well. Then, once you have the experiment, you can go through and set up the Lean Analytics so that you can measure your KPI and carry on in the experiment.

### Measure your outcomes and make a decision

You can't just get started with an experiment and then walk away from it. You need to measure how well it goes to determine if it is truly working; if some changes are needed; or if you need to work from scratch. You can then make a decision on the next steps you need to take. Some of the things to look for when measuring the outcomes during this stage include:

- **Was the experiment a success?** If it is, then the metric is done. You can move on to finding the next metric to help your business.

- **Did the experiment fail?** Then it is time to revise the hypothesis. You should stop and take some time to figure out why the experiment failed so that you have a better chance at a good hypothesis the next time.

- **The experiment moved but was not close to the defined goal.** In this scenario, you will still need to define brand new experiment. You can stay with the hypothesis if it still seems viable, but you would need to change up the experiment.

# Chapter 4: False Metrics vs. Meaningful Metrics

One of the prerequisites for working with Lean Analytics is understanding that most people are using their data wrong. When you don't use your data correctly, you are not going to be able to come up with patterns, opportunities, or results that are achievable.

There are two points that come with this idea. These two points are:

- There are many companies, as well as people, who will label themselves with descriptions like "data-driven." Sure, they may use up a lot of their resources on compiling data. However, they then miss out on the "driven" part. Few are actually going to base a strategy on the information that they acquire from the data. They may have the right data, but they either don't understand it or choose to react to it incorrectly.

- Even if the actions of a company or person are driven by data, the problem of using wrong data still exists. Often, they will oversimplify these metrics and then use them according to the convention. Keep this in mind: just because other people do this or have done this doesn't mean it is going to prove useful to the goals that you have. Consequently, the data is going to become garbage in, and then the analysis is garbage out. This is often known as false data.

As a business who is interested in working with Lean Analytics, it is important to learn the difference between false metrics and meaningful metrics. If you follow false metrics, you are going to be following a strategy that is not going to help you reach your goals, which will mean a lot of time, effort, and resources wasted.

## *The biggest false metrics to watch out for*

As a business that is trying to cut out waste and ensure that you provide the best customer service and the best products possible, you must always ensure that you watch out for some of the false metrics that may come up. Many people who don't understand how data works will be taken in by these false metrics that, in reality, will mean wasted time and resources. Some of the most common false metrics for you to watch out for include:

- **The number of hits:** Just because you have a website that is attractive and contains many points of interest, doesn't necessarily mean that it will tell you what the users are really interested in. You should not focus on the number of hits your website gets. This may make you feel good about your website, and it can be neat to see how many people come and visit your website. But you need to focus more on what the customer is interested in or looking for.

- **Page views:** This metric refers to how many pages are clicked on a site during a given time. This is a slightly better than hits, but you typically don't want to waste your time with this metric. In most cases, unless you are working with a business that does depend on page views, such as advertising, the better metric for you to use is to count people. You can do this with tools that will provide information on unique visitors per month.

- **Number of visitors:** The biggest problem with this metric is that it is often too broad. Does this type of metric talk about one person who visited the same site a hundred times, or a hundred people who visited once? You most likely want to look at the second group of people because you've obtained more impressions. Otherwise, just looking at the number of visitors will not give you this information.

- **Number of unique visitors:** This is a metric that is going to tell you how many people got to your website and saw the home page. This may sound good at first, but it is not going to give you any valuable information. You may also want to find out things like how many visitors left right away when they saw the page or how many stayed and looked around. Unique visitors can help you see that some new people are coming onto your website and checking things out but they don't really tell you much about those visitors or what they are doing.

- **Number of likes, followers, or friends**: This is a good example of a vanity metric that is going to show you some false popularity. A better metric that you can go with is the level of influence that you have. What this means is how many people who will do what you want them to do. While it is good to have followers and likes on your page to show that people are looking at your content, it is not as important as some of the other metrics that you can pick.

- **Email addresses:** Having a big list of email addresses

is not a bad thing by itself. But just because you have this large list does not mean that everyone on it is going to open, read, and act on the messages that you send out. You want to make sure that the email addresses that you do have are high quality and are from people who actually want to hear from you, even if that means your email list is a little bit smaller. If you are collecting emails, strive to get addresses from people who are actually interested in your product and services. Don't just collect emails so you can boast of a large list.

- **The number of downloads:** This is a common metric that is used for downloadable products. While it can help with your rankings in the marketplace when you are in the app store, the download number is not going to tell you anything in depth, and it won't give you any real value. If you would like to get some precise answers here, you can pick some better metrics. The Launch Rate is a good place to start because it will show the percentage of those who downloaded, created an account, and then used the product. You can also use something like Percentage of Users Who Pay so you can see how many actually pay for anything.

- **Time spent by customers on a page or website:** The only time that this is going to be useful is for businesses that are tied directly to the behavior of the engaged time. For example, a customer could spend a lot of time on your web page, but what if they are spending that time on the help pages or on the complaints pages? This is not necessarily a good thing for your business, so this metric is not the best one for you to go with.

If you want to pick out a metric that will actually help your business get ahead, then you must make sure that you avoid some of these false metrics. They may look good on the surface, but in reality, they are just giving you information that could be pretty useless, and you will end up wasting a lot of time and money to follow them.

# Chapter 5: Recognizing and Choosing a Good Metric

Part of the Lean Analytics methodology is finding a good metric to help you out. The Lean Analytics Cycle is a measurement of movement towards a goal that you already defined. So, once you have taken the time to define your business goals, then you must also think about the measurements you can make to progress towards the goals.

This can be hard to do. How are you supposed to find a good metric that can make sure you go towards the goals that you set out? Some of the characteristics that you can look for when searching for a good metric include:

- **Comparable:** You know that you have a metric that is good if it is comparable. You want to be able to compare how things have changed in the last year, or even from one month to another. This gives you a good idea if there have been any changes, positive or negative, with your business process, customer satisfaction, and more. You can ask yourself these questions about the metric to help test for this:

  o How was the metric last year, or even last month?

  o Is the rate of conversion increasing? You can use the Cohort Analysis to help with conversion rate tracking.

- **Understandable:** The metric that you use should never be complex or complicated. Everyone should be able to understand what it is. This ensures that they know what the metric is measuring.

- **Ratio:** You should never work with absolute numbers when you are working with metrics. If you find that you have those, you should try to convert them to make comparisons easier, which in turn makes it easier to make decisions.

- **Adaptability:** If you have chosen a good metric, it should change the way that the business changes. If you notice that the metric is moving, but you have no idea why it is moving, then it is never a good metric. The metric should move with you, not randomly on its own, or it won't be a secure one to use.

## *Types of Metrics*

There are two metric types that you are able to use when doing Lean Analytics. These include qualitative and quantitative metrics. To start, qualitative means that the metric has a direct contact with your customers. This would be things such as feedback and interviews. It is going to provide you with some detailed knowledge of the metric.

You can also work with quantitative metrics. These are more of a number form of metrics. You can use these to ask the right types of questions from the customer.

Of course, both of these methods have other things under them that make them easier to use. You will find that both of these methods have actionable and vanity metrics.

- **Vanity metrics** will not end up changing the behavior of the thing you are concerned about. These are a big waste of your time, and you should avoid them as much as possible. They seem to provide you with some good advice and something that you can act upon, but often they don't lead you anywhere and can make things more difficult. If you are working with a company to help you determine your metrics, be very wary if they start touting the benefits of following any of the vanity metrics.

- **Actionable metrics** are going to end up changing the behavior of the thing you are concerned about. These are the types of metrics that you want to work with on your project. They are metrics that can lead you to the plan that you should follow and can make it easier to come up with a strategy to make your business more efficient.

- **Reporting metrics** is a good way to find out how well the business is performing when it does even everyday activities.

- **Exploratory metrics** are going to be useful for helping you to find out any facts that you do not know about the business.

- **Lagging metrics** are good to work with when you want more of a history of the organization and you want as many details as possible to help with a decision. The churn of a company can be a good example of the lagging metrics. This is because it is going to show you how many customers have canceled their orders for a specific amount of time.

79

- **Leading metrics** are good because they can help provide you with the information that you need to make future forecasts for the business. Customer complaints can be a good example of leading metrics because it can help you to predict how a customer will react.

You will need to determine which kind of metric you want to use based on the problem or project that you are working on. Working with one metric is usually best. Doing so will help keep you on track, so you know what to look for. Don't waste your time trying to work on more than one metric. You will only get confused and end up with no clear idea about the strategy to follow.

# Chapter 6: Simple Analytical Tests to Use

Another thing that you should concentrate on to do well with Lean Analytics is to have some familiarity with the tests that are used. These tests are helpful because they are going to be used to help you examine any assumptions that you are trying to use here. These tools can also be used to help you identify customer feedback so you can respond to them properly. Let us take a look at some of the best analytical tests that you can use when working with Lean Analytics.

## *Segmentation*

The first test is segmentation. This process involves comparing a set of data from a demographic bucket. You can divide up the demographics in any manner that you choose such as gender, lifestyle, age, or where they live. You can use this information to find out where people are purchasing a particular product; if there are different buying behaviors between female and male customers; and if your target audience seems to be in a certain age group or not.

The reason that you want to build up a user segment is to make it easier for the data to be actionable. Analytics can teach you a ton about the people who purchase from you, but there is often a lot of information there, and it can be hard to draw good conclusions from this information. After all, while this information from the past can be useful, it isn't going to be the best to tell you how to improve either retention or conversion rate.

This is where the process of segmentation is going to come into play. When you learn how to filter out the audience, you will then be better able to create a plan to make new products that serve them the most. Analytics can give you the information that you need, but segmentation can help you to act.

For example, you may have a conversion rate that seems average or good, but it could be from a combination of one group that converts really high and consistently so, and then another group that seems to never convert at all. You could be wasting a lot of money on that second group where you are hardly getting anybody to convert at all. Segmentation can be used to help you understand what things you are doing the right way when engaging the first group, and can give you a plan on how you can improve to work on that second group.

With segmentation, you don't want to only look at the data to learn some more about your users, but you also want to come up with data that you can act upon. Segmentation can help you with this. You will be able to divide up the people in your customer base and learn how to advertise to them better than ever before.

Remember that not all customers are going to be the same. There are some of your customers who may purchase something once, and they aren't regular customers. While it is still good to reach out to them, you want to learn who your regular audience is, what they respond to, and what keeps them coming back. This is going to ensure that you keep them coming back and earn as much profit as possible.

So, how do you create a segment of your customers? There are many different options that you can use when creating a segmentation. But let's look at the process that you can use to create a segmentation for your Lean Analytics project. The steps you need to use include:

- **Define the purpose of your segmentation:** You should first figure out how you want to use your segmentation. Do you want to use it to get more customers? Do you want to use it to manage a portfolio for your current customers? Do you want to reduce waste, become more efficient, or something else? Defining your purpose can make it easier to know how you should segment out your customers.

- **Identify the variables are the most critical:** These are going to help influence the purpose of your segmentation. Make sure that you list them out in order of their importance, and you can use options like a Decision tree or Clustering to help. For example, if you want to do a segmentation of products to find out which ones are the most profitable, you would have parameters that are revenue and cost.

- Once you have your variables, you will need to **identify the threshold and granularity of creating these segments.** These should have about two to three levels with each variable identified. But sometimes you will need to adapt based on the complexity of the problem you are trying to solve.

- **Assign customers to each of the cells.** You can then see if there is a fair distribution for them. If you don't see this, you can look for the reasons why, or you can tweak the thresholds to make it work. You can perform these steps again until you get a distribution that is fair.

- **Include this new segmentation in the analysis**

and then take some time to look it over at the segment level.

## Cohort Analysis

The Cohort Analysis is a test involves comparing sets of data using a time bucket. In this test, there will be differences in behavior between customers who arrived at the free trial stage of your process, versus those who showed up at the initial launch, and then those who are in the full payment stage.

Each of these is significant because it helps you to figure out which customers are likely to come back and be full-fledged customers when in the future. Those that show up in the initial stages when the product is free are often not the customers you are going to see when sales actually start. They may have just wanted to try it out and didn't really have an investment in the product.

Those that are in the later two stages can be customers who are better for you to work with. They will be the most interested in the product because they invested some money to get it. You really want to study these using the cohort analysis to figure out who your real customer base is and how they behave so that you can better market to them later on.

## A/B Tests

A/B testing is a process where you examine an attribute between two choices. This could be something like an image, slogan, or color so that you can figure out which option is the most effective choice.

Let's say that you had two products that you are comparing and you want to find out which ones customers liked the best. Did they choose one product over the other and why? Did they respond better to the choice that was in green or the one in blue?

For this test to really work, you must assume that everything else is going to stay the same. So, it would have to be the same product, but there is one variable that is different between them. You could put up a website, for example, and have a red background on one version and a yellow background on another. Then you could use A/B testing to figure out which one the customer responded to the best out of those choices.

In addition, you can also work on multivariate analysis. This is pretty much the same thing, but instead of going through and testing out one attribute, you will go through and compare several changes against another group of changes to see which is the most effective. This one will require there to be a few changes in the second product compared to the first to be the most effective.

There are several keys that you need to have in place when you are ready to do an A/B test. These include:

- Know the reason that you are running this A/B test.

- The item that you are testing needs to be noticeable to the audience. If you make a minor change that no one is going to notice, then your results are not going to be that reliable.

- Stick with testing just one variable at a time. If you go through and do multivariate testing, or test more than one thing at a time, you will run into trouble. You may not know for sure which variable is causing the changes you see.

85

- Your test needs to end up being statistically significant. This means that it must have a sample size that is big enough to test and know that the results are valid within a certain margin of error.

Let's take a look at an example of how to do this. We are going to use this test on a website that you are trying to improve. There are two main ways that you can do this including:

- You will test the pages on separate pages.

- You will use JavaScript to conduct the test inside the page, so you don't need two different URLs to do it.

The first option is going to mean that you will need to have two different URLs for the pages that you are testing out. You can make them similar names, but make sure that there is some way that you can keep track of them and not get the two confused.

With the second option, you will need to have some experience working with JavaScript. You can then place some of this code on the website so that it can dynamically serve one option or the other.

The method that you choose is often going to depend on the one that you like the most and which tools you want to use. Both of these will give you some valid results, but you will find that implementing each of them takes a different amount of time to set it up.

# Chapter 7: Step 1 of the Lean Analytical Process: Understanding Your Project Type

Now that we have taken a look at some of the different part of Lean Analytics, it is time to take a closer look at how the process works. These can help you to get started with the Lean Analytics stage for your business and ensure that you are getting the most out of Lean.

The first step that we are going to look at is understanding your business or your project type. How are you supposed to pick out the right metrics if you have no idea what kind of business or project type you are working on? You must really understand the project at hand so that you can choose a fantastic metric that can show you results.

There are six general business types that you can fit into, and they all will have metrics that are going to work best or matter the most, for each one. If you see that your business or project is on this list, your job will be simple. You just need to focus your attention towards understanding the priorities of what needs to be measured. This can include in-depth external research.

However, if you have a business that is not on this list, this doesn't mean you are out of luck and can't do anything. You can just use some of the information that is in this chapter as an example and build up your own understanding and metrics from this chapter.

## E-commerce

The first type of business is going to be e-commerce. These are growing like crazy right now as many customers are looking for the things they want to buy online more and more. And many companies find that they can make large profits by offering their products and services online to these customers. An e-commerce business is going to be any that has their customers buy from a web-based store. This could include businesses such as Expedia.com, Walmart.com, and more.

The strategy for this type of business is that you need to understand the customer relationship that you want. This means that you are going to focus either on new customer acquisition or customer loyalty? You have to decide between these two because this is going to help with all other decisions that you make with this type of business.

There are many metrics that you can choose to go with in an e-commerce business. Some of the typical ones that other companies have chosen in this industry include:

- Inventory availability
- Shipping time
- Mailing list and how effective it is
- Virality
- Search effectiveness
- Shopping cart abandonment
- Revenue that you make on each customer
- The amount you spend to get new customers
- Shopping cart size
- Repeat purchase

- Conversion rate

## The best metrics

Of course, there are several metrics that will work the best and will provide you with the best return on investment, when working with an e-commerce site. The best metrics to use here include:

- **Conversion rate:** This is the percent of all visitors to your site who also purchase something. The average conversion when it comes to online retail is 2%. There are some that can do better though. For example, Tickets.com is over 11%, and Amazon.com is at almost 10%.
- **Shopping cart abandonment:** It is typical that 65% of the shoppers to your website are going to abandon their carts. Many of these are because of the high costs of shipping, and others are from the high price of all the items in their cart. You should definitely take some time to analyze any shopping cart abandonment that is happening in your business so that you can learn why you are losing these customers.
- **Search effectiveness:** The majority of your buyers are going to have to search to find what they need. If you make your search more effective, it can help your customers find what they want, rather than having them leave in frustration. Remember that about 79% of your total shoppers will use the search engine for half of the goods they want.

## *Software as a service*

These types of companies are going to sell software in downloadable form or as a subscription. This can be things such as Skype, Evernote, Basecamp, Adobe, and more. They are not selling a physical product to someone, but these software programs are still pretty important for most people to get work done or to do other things on their computers.

The strategy with this one is that most software is going to consist of products that are on a subscription which means that retaining the customers is important. Your success is going to really depend on building up a loyal base of customers faster than those customers disappear.

There are some metrics that you can use to make this happen. Some of the most common metrics that are used with this type of business model include:

- Reliability and uptime
- Upselling
- Virality
- Customer churn
- Customer lifetime value
- Cost of getting new customers
- The amount of profit you make per customer
- User conversion
- User stickiness
- User enrollment
- User attention

## The best metrics

Just like last time, you are able to use any of the metrics that are above, but there are some that could be the best for helping you reach your overall goals. Some of the best metrics to use with a software company includes:

- **Paid vs. free enrollment:** You will find that your enrollment rate is going to change depending on whether or not you asked for credit card information in the free stage or not. The former is going to get an average signup rate of 2 percent, and then 50 percent often end up buying. When you do not ask, the average may increase to ten percent, but only 25 percent purchase the product.

- **Growing revenues and upselling:** Some of the best software providers are able to get 2 percent of their paying subscribers to increase what they pay each month. Being able to grow your customer revenue by 20 percent in a year can be achieved if you work towards it.

- **Churn or attrition rate:** This is the percent of your customers who are leaving. Going across the industry, the top companies usually have an attrition rate between 1.5 and 3 percent each month. If you have a percentage that is higher, then you need to find ways to make the customers stay.

### *Mobile app companies*

These are companies that are going to provide apps to be used on mobile devices like Android and iPhone. Some of the companies that can fall under this category would be ones like

WhatsApp and Instagram.

The strategy that you want to go with here is to find the right target audience. There are a lot of ways for your app to make money, but you will find that the majority of your revenue is going to come from a smaller group of customers, rather than from the population as a whole. You should focus your analysis as well as the metrics you use to help you better understand those customers.

There are many metrics that you are able to use as an app company. Some of the most common options include:

- Customer lifetime value

- Churn rate

- Ratings click-through

- Virality

- The revenue you make from each paying user

- The revenue you make for each user

- Percentage of users who end up paying

- How much it costs to get the customers

- Launch rate

- Downloads

## Best metrics

Of course, there are many metrics that you can choose to look at when it comes to being an app company, but a few of them are going to provide you with the most information and can help your business to really grow. Some of the best metrics you can use include:

- **Downloads and the app launches:** The number of people who download the product and then activate it will fit in here. It is known that quite a few people who decide to download an app will then never activate it or use it at all, especially if the app is free.

- **The cost to get new customers:** You can follow a general rule to have a budget of 75 cents per user in your marketing initiatives to help attract new customers. You should always make sure that the cost to get new customers is lower than what you will earn on them. So, if you will only earn 50 cents on a customer, then you shouldn't spend 75 cents on each one.

- **The average revenue you earn per customer:** This is often going to be determined through the business model. For example, Freemium apps, or apps that you receive revenue from engagement in the app, will often have a higher revenue per user compared to those that are premium apps.

### *Media site companies*

If you are in this industry, you have a website that is going to provide some information, such as articles, in return for earning advertising or any other type of revenue. These would include most blogs and other sites like CNN.com, CNET, and more.

Media sites need to really understand the source of their revenue. It is not coming directly from their readers or the people who use their "product," but it is coming from advertisers who are trying to reach those readers. So, if you are a media site company, you would get revenue from affiliates, click-based advertising, display advertising, and sponsorship. You would want to design your key metrics to work for this.

Some of the different metrics that you can choose to work with for a media site company include:

- Page inventory

- Pages per visit

- New visitors

- Unique visitors

- Content and advertising balance (you don't want too much advertising on the page, or it takes away from the content and keeps the customer away).

- Click through rates

- Ad rates

- Ad inventory

- Audience and churn

**Best metrics**

- **Click through rate:** This is the number of users that are going to click on a link out of all the users who check out the page. The average click-through rate for a paid search in 2010 is 2 percent, but some companies can get higher. If you see that you are at one percent, then it is time to make some changes. But if you are above that number, you are doing really well.
- **Engaged time:** This is how long your reader will stay on the site and look through the content and the ads. Most media sites are going to aim for 90 seconds for content pages, and a little less with landing pages. If you find that your visitor is not spending more than a minute on the content pages, then it is likely your content is not engaging them.
- **Content optimization for media:** This one means taking the content that you already have and changing it so that it works on other venues, such as podcasts and video. You should track how others are using the materials you have because this can help you find some new opportunities to use.

### User-generated content business

If you have a community that is engaged, they are going to contribute free content. And this same engagement is going to provide you with ads as well as other revenue sources. Some examples of companies that work with this include forums, Wikipedia.com, Reddit.com, Facebook.com, and Yelp.com.

The strategy that you should use is one that takes into account user engagement. This business is going to be successful when its visitors become regular contributes, and they interact with others in the community and provide quality content. User engagement tiers to measure involvement can be good as well.

Some of the different metrics that you may want to use with user-generated content include:

- Notification and mail effectiveness

- Content sharing

- Value of the content that is created

- Content creation

- Engagement funnel changes

- How many engaged visitors you have

**The best metrics**

- **Time on the site each day**: Here you are going to measure how long the average user is on your site and engaged on a typical day. This is a good thing to measure for engagement and stickiness. The average number is about 17 minutes a day, though Facebook is usually an hour, and Tumblr and Reddit are 21 and 17 minutes respectively.
- **Spam/Bad Content:** With these kinds of communities, you need to make sure that good content is always uploaded. You will have to spend time and money to keep bad content and fraudulent content off the site. You can measure what you think is good and

bad and then build up a system to help keep up with this. You can also spend your time watching out for quality decline and then fix it before it ends up ruining your community.

## *Two-sided marketplace business*

These kinds of businesses are going to connect buyers and sellers, and they will earn a commission on the work. It is kind of a variation of the e-commerce store. Some options of this would include Priceline.com, Airbnb.com, Ebay.com, and Etsy.com.

The strategy with this business is that you need to be able to attract in two different customers, the buyers and the sellers. The best bet is to focus on those that have the money to spend first. If you can find a group of people who want to spend their money, then those who want to make money will pretty much line up to do it.

Some of the metrics that you are able to use when it comes to a two-sided marketplace business include:

- The volume of sales and the revenue you earn

- Pricing metrics

- Ratings and any signs of fraud showing up

- Conversion Funnels

- Search effectiveness

- Inventory growth

- Buyer and seller growth

**Best metrics**

- **Transaction side:** Sellers usually won't have the money or time to analyze pricing and the effectiveness of their copy and pictures. As the owner, you will have the aggregate data from all your sellers, and you can use this information to help them with this analysis. Transaction size is the same as the purchase size, and of course, it is going to differ based on your business type. You should help your sellers measure it so that they can understand the behavior of your buyer and use it to sell more items.

- **Top 10 lists:** You can make top ten lists to help your buyers find the best products, and your sellers to know what is going on in the industry and what they can do to be more profitable.

As you can see, there are many different types of businesses out there. And it is likely that your business is going to fit somewhere in this list. If it does, then there is an outline that you can use for developing a good strategy. Even if you don't, you can combine a few of these strategies to help you come up with the metrics, and the plan, that you need to succeed.

# Chapter 8: Step 2: Determine Your Current State

Now that you know which business type you are in, it is time to move on to the second step of Lean Analytics. This one is going to require you to determine which innovation stage you are in right now. The one metric that means the most to you right now is a function of time. It is going to change as your project keeps moving on through the different stages of innovation. There are several different stages of innovation that you can work with. These include:

### Stage 1: Empathy or is this a real problem?

In this stage, you are going to identify a problem in your business and then get inside the head of your potential user. You should be in their shoes and understand why there is a problem and what they are thinking. You may need to spend some time talking to potential customers to help with this stage. The more that you are able to talk to your potential customers and others in the market about the product or service you want to offer, the better off you will be. This can give you some real insights that can drive your business forward.

You need to focus on any metrics that are going to help you to determine whether or not the problem is harmful to your business. The metric needs to also determine if there are enough people who care about this problem. If only a few people see it as a problem, then it probably isn't worth your time taking care of it. But if a big percentage sees this as a problem, then it is something to take care of. You can also use metrics that will see what the success rates of your existing solutions are and if you need to change some of them.

### Stage 2: Stickiness or do I have a good solution?

In this stage, you are going to start by making a Lean prototype of your solution to the problem you found in the previous step. You have to ask yourself whether or not people will pay for this. This is when you can gain feedback from small focus groups and testers. Based on that information, you can make adjustments and changes to the solution until you get it right.

You are going to need to focus on any metric that proves your solution will encourage the user to engage and also come back to your business.

### Stage 3: Virality or does this solution provide value to enough people?

Once you have a solution and a product and they are seen as effective, you need to decide whether its value adds enough that the customer will tell others about it. Remember that word of mouth endorsements are valuable as a precursor to growth measurement and as free advertisement for the business. You can work for endorsements that are either natural (the customer enjoys the product or service enough that they just give out recommendations to their friends) or ones that are incentive-based (such as giving the product for free or at a discount).

For this one, you need to focus on metrics that are able to measure out if you are getting any new customers from your existing ones. And you want to know how many of these referrals are happening. You can also take a look at metrics that can check for how long it takes for news to spread or the cycle time.

## Stage 4: Revenue or can I make this profitable?

Now you need to work on how much revenue you can expect from selling the product or service that you created in the last step. You can work on prices, standardization, control costs, and margins. You need to take the time in this step to prove that you are able to make money in a self-sustaining and scalable way, or this is not the solution for you.

This one is going to need you to focus on metrics that can tell you the net revenue that you are able to earn for each customer. The net is going to be the revenue that you make per customer minus the amount you spent to get that customer.

## Stage 5: Scale or can we expand to a bigger audience?

Now that you have a product, you showed that it is effective, and you have a business model in place to show that it is going to be functional and profitable, you can now invest and expand it into new markets. This can include new geographies, channels, and audiences as well.

If you are dealing with a project or business that is oriented on efficiency, you need to focus on metrics that are able to reduce costs. If you are working in a business or project that is differentiation oriented, you will want to focus on metrics that will track margins for you.

### What can I do with these innovation stages?

Now that you know a little bit more about these innovation stages, it is time to figure out where you are and learn what you can do with each one. The steps that you should take from here include:

- Look through the stages above and determine where your business or project is right now.

- Refocus on the things that you should be measuring at the stage you are in.

- If you find that your project does not fit into this framework, then it is important to remember that all innovative endeavors are going to follow a pattern of stages as well as maturity through time. Are you able to borrow this framework and leverage it in some manner so that you can figure out what stage your project is in right now? This can really make a difference in helping you to understand what you really need to be focusing on right now.

Knowing where you are in the innovative stage can make a big difference. When you look at the five stages above, you will have a clear outline of what you need to focus on and what needs to be done to keep you moving forward. If you have no idea where you are right now, then how are you supposed to know what steps to take to get to the next level? Always have a good idea of where you are in the innovation process, and then you have a clear picture of where you should go next and can keep on track.

# Chapter 9: Step 3: Pinpoint the Most Pressing Metric

To help you to be successful with any type of innovative project, the key is to focus. Consequently, you can't spend your time on too many metrics because this is going to make you feel distracted and you are going to lose all your focus.

In the first few stages of innovation, it is often best to reduce the number of metrics that you track. If you can, you should focus on just one metric, the one that matters the most right at the moment. The metric that is the highest-priority is usually related to the most important project or business need.

For example, a subscription software company may be in the virality stage, and they are trying to gain traction with it. They may decide that the net adds metric is the one that will help them out the most. Remember that Net Adds = Total of New Paid Subscribers – Total That Cancelled.

There may be other metrics that your company can use, but you need to just focus on one. You will need to figure out what problem is the most pressing or the most important right now, and then go with a metric that fits with this the most.

### *How can I find that one metric?*

Some of the steps that you can use to find the one metric that matters the most right now include:

- Write down the top three to five metrics that you really like, and you often track.

- How many of these metrics are actually any good and help you out?

- How many do you use to make business decisions? How many of those would actually be vanity or false metrics?

- What stage are you at with the business or the project? Do you really understand what matters in the business model? Can you discard any of the metrics that aren't really adding value to you right now?

- Are there any other metrics that are not on your list that you can think of and that you think could be more useful right now?

- Once you have written down all those metrics, you can go through the list and cross off any bad or false metrics. Add any new good ones that you think of on the bottom.

- Now that you have a list, you should go through and pick the one metric that you absolutely can't live without to help you with the project in its current stage.

### *What to do after optimizing that one metric*

Once you have the metric and the project at a level where you are happy with the numbers, you must remember that you will need to continue measuring them. You never know when that project or that metric will need to be changed up again to help you in the future. But you can rest assured knowing that the process is now controlled and optimized. What this means is that you are now at a point where you are achieving a certain level of results.

Now you are able to go back to your list and work on the metric that is the next highest priority. This is going to be the next highest priority of your business, or the next biggest project need. You can review through the innovation stage you are currently at and the business or project type, and then you can determine which point of interest you should focus on.

Remember that the goal here is to only work on one metric at a time. There may be several metrics that need to be addressed in your business. Keep your focus on one at a time.

Sure, you can go through and write out a list of the different things that need to be addressed at some point, and Lean Analytics is a good time for this because it can help you see what problems are there. But you should pick the one that is the most pressing and work on that one first.

After you have time to complete the Lean Analytics stage on this problem and you have a winning strategy in place for it, then you can move on to the next step of picking out a new project to work on. You can implement this process on as many projects as you would like. Just make sure that you are only working on one at a time.

# Chapter 10: Tips to Make Lean Analytics More Successful for You

Getting started with Lean Analytics is something that can take some time to get used to. It is going to provide you with great results and a winning strategy that is sure to get you ahead. But for those who are just starting out with this stage, or who are just getting started with the whole idea of the Lean system, you may need some help to get going on the right foot. Here are some great tips that you can follow to ensure that you are doing well with Lean Analytics and to ensure it is as successful as possible for you:

- **If you are doing an A/B test, you need a lot of users:** You are not going to get any good results from you A/B test if you don't have a lot of users to help you out. This means that it is not going to work all that well if you are a small startup or if there are not a lot of people you can measure. You should have a minimum of 10,000 events before you attempt this kind of test. These events can include visits or people who use a feature. Make sure that you are able to get this many users to help you out before you get started.

- **Make big changes:** If you are not able to see the changes from a few feet away, it is likely that the people you are testing with the A/B test won't either. For example, A/B tested 41 different shades of blue. The results were not the best because there were just too many different shades and for most people, they looked too similar. You need to make big changes before you do an A/B test, or it won't work well for you.

- **Measure the tests properly:** You are not going to get the right results if you are not properly measuring the tests. You need to have the right metric in place. Also, make sure that you never stop a running test too early, or you may miss out on some of the important results that you need.

- **Use the tools that you need:** Lean Analytics has a ton of tools that you can use to make it successful. Make sure that you are properly trained to handle each part and that you don't miss out on some important tools that can make this more successful.

- **Know where your business is now:** How are you supposed to have any idea of what kind of project to work on and what metrics to use if you don't have a good understanding of your business? Make sure that you know the overall goals and vision of the business. This can help you to spot some of the problems that you need to fix and can make it easier to ensure that whatever changes you do decide to make are going to go along with what your business is all about.

- **Understand the different metrics:** You should spend some time looking at the different metrics that are available for you to use on your project. Each one can be great, but it does depend on the type of business that you are running and the project that you want to work with. You need to learn which metric is going to be the right one for you.

- **Add this into the Lean Support System:** Many people are fond of the Lean Support System. This allows them to get rid of a lot of the waste that their company may have, and can make them more efficient. But you need to do Lean Analytics first to see success. This helps you to gather the information that is needed and then sort through it and analyze it. Then you can use this information to come up with the best plan to handle your problem. If you just jump into a strategy without the resource, it is likely that you won't see results at all.

- **Focus on the main problem first:** If you are like many businesses, there are probably many problems that you need to solve. But you don't have the time and resources to do all of them at the same time. When you get started with Lean Analytics, you must figure out what the main issue is, the one that will have the biggest impact on your profits, and work with that first. Once you have successfully implemented Lean Analytics and worked on the problem, then you can go back and see if there are any other problems that need to be addressed.

- **Get rid of the waste:** Remember that the most important thing that you will do with the Lean system is get rid of waste. And the data that you collect in the Lean Analytics stage is meant to help you to find the waste and learn how to get rid of it. Take a look at some of the most common types of waste that businesses may experience (and that we listed in an earlier chapter) to give you a good idea of where to start.

Lean Analytics can be a great way for you to get a strategy together that will help your business become more successful. If you follow these tips and some of the strategies that we talk about in this guidebook, you are sure to see some amazing results in no time.

# Conclusion

Thank you for making it through to the end of *Lean Analytics: The Complete Guide to Using Data to Track, Optimize and Build a Better and Faster Startup Business*, let's hope it was informative and able to provide you with all of the tools you need to achieve your goals.

The next step is to start the process of implementing Lean Analytics into your own business. Learning how to make changes so that you can be more cost effective and provide better service to your customers all starts with Lean Analytics. This stage asks you to search for the data you need and analyze it so you know what step to take next. You can't come up with a plan for improving your business without the help of Lean Analytics to make it possible.

Finally, if you found this book useful in any way, a review on Amazon is always appreciated!

# Lean Enterprise

*The Complete Step-by-Step Startup Guide to Building a Lean Business Using Six Sigma, Kanban & 5s Methodologies*

# Introduction

Congratulations on getting a copy of *Lean Enterprise: The Complete Step-by-Step Startup Guide to Building a Lean Business Using Six Sigma, Kanban & 5s Methodologies.* These days, it is more difficult than ever to build a business that can remain competitive in a world where customers can find your competition with just the click of a mouse. While there is only so much you can do when it comes to adjusting your profit margins, you can still find success by adjusting the method that will complete the processes in making your business successful.

Making a business Lean can give it the competitive advantage that the perpetual buyers' market takes away. However, it can be more difficult than it first appears which is why the following chapters will discuss everything you need to know in order to turn your business into a Lean mean fighting machine. First, you will learn all about the Lean system, its many benefits, and how you can get started creating your very own Lean system. Next, you will learn how to move the process forward in the right way by ensuring that you have the right goals in mind and that you go about implementing them in the best way possible.

From there, you will learn how to create a value stream map which is vital when it comes to ensuring that your business's various processes are truly on point before learning how to choose the Lean system that best supports the flow of production. You will then learn about the importance of standardization before learning about the several important Lean tools which you can use to really whip your business into shape.

There are plenty of books on this subject on the market, thanks again for choosing this one! Every effort was made to ensure it is full of as much useful information as possible, please enjoy!

# Chapter 1: Why Lean Matters to Your Enterprise

Lean principles were first discussed by an MIT student by the name of John Krafcik in his master's thesis. Before starting at MIT, Krafcik had already spent time as an engineer with both Toyota and GM, and he used what he learned from the Japanese manufacturing sector to posit a number of standards that he believed could help businesses of all shapes and sizes operate more efficiently.

The basic idea is that, regardless of what type of business a business is in, it is still just a group of interconnected processes. These interconnected processes can be categorized such as primary processes and secondary processes. The primary processes are those that directly create value for the business. Meanwhile, secondary processes are necessary to ensure the primary processes run smoothly. Regardless of the type of process you look at, you will find that they are all made up of a number of steps that can be carried out in a way that ensures they work as effectively as possible and that they need to be viewed as a whole in order for an effective analysis to be completed.

As a whole, you can think of the Lean process as a group of useful tools that can be called upon to identify waste in the current paradigm either for the business as a whole or for its upcoming projects. Specific focus is also given to reducing costs and improving production whenever possible. This can be accomplished by identifying individual steps and then considering the ways they can be completed more effectively. Some tools that are commonly used in the process include:

- 5S value stream mapping
- error-proofing
- elimination of time batching
- restructuring of working cells
- control charts
- rank order clustering
- multi-process handling
- total productive maintenance
- mixed model processing
- single point scheduling
- single-minute exchange of die or SMED
- pull systems

Beyond these tools, Lean is also comprised of a number of principles that are loosely-connected around the twin ideas of the elimination of waste and the reduction of costs as much as possible. These include:

- flexibility
- automation
- visual control
- production flow
- continuous improvement
- load leveling
- waste minimization
- reliable quality and pull processing
- building relationships with suppliers

When used correctly, these principles will ultimately result in a dramatic increase in profitability. When given the

114

opportunity, the Lean process strives to ensure the required items get to the required space in the required period of time. More importantly, however, it also works to ensure the ideal amount of items move as needed in order to achieve a stable workflow that can be altered as needed without creating excess waste.

This is most typically achieved via the tools listed above but still requires extreme buy-in at all levels of your organization if you ever hope for it to be effective in both the short and the long-term. Ultimately, the Lean system is only going to be as strong as the tools your company is using to implement it, and these tools will only ever be effective in situations where its values are expressed and understood.

## Important principles

While it was originally developed with a focus on production and manufacturing, Lean proved to be so effective that it has since been adapted for use in virtually every type of business. Before adopting the Lean process, businesses have only two primary tenants. The first one focuses on the importance of incremental improvement while the other one is the respect for people both external and internal.

*Incremental improvement:* The idea behind the importance of continuous improvement is based on three principles. The first is known as the Genchi Genbutsu and is discussed in detail below. The second is known as Kaizen, and it has its own chapter later in this book. Finally, in order for continuous improvement to be truly effective, it is important to understand that you must lead your business with a clear knowledge of the challenges you are most likely to face as it is the only way to determine how to deal with them effectively.

When doing so, it is important to approach each challenge with the appropriate mindset which is one that supports the idea that every challenge leads to growth, which, in turn, leads to positive progress. Finally, you will also want to ensure that you take the time to challenge your preconceptions regularly as you will never know when your business might end up operating on an assumption that is no longer true. This is ultimately the best way to find unexpected waste which will ensure that you really start to improve internally not just in the short-term but in the long-term as well.

*Respect for people:* This tenant is both internal and external as it applies equally to your own people as well as to your customers. Respecting the customers means going the extra mile when it comes to considering their problems and listening to what they have to say. When it comes to respecting your team, a strong internal culture that is dedicated to the idea of teamwork is a must. This should further express itself in an implied commitment to improving the team as a means of improving the company as a whole.

## Getting an edge

Prior to the digital age, businesses could determine their sales margin by starting with all the relevant costs, adding on a reasonable profit margin and calling it a day. Unfortunately, the prevalence of screens in today's society means that everyone is a bargain shopper, simply because it takes so little effort. This, in turn, means that you are not only competing against other businesses in your city, county or state, you are now competing with businesses all around the world as well. As such, there are only a few options when it comes to squeaking by with any profit margin whatsoever. Companies can either add additional real or perceived value, or they can reduce the amount of waste they are paying for as much as possible.

Most businesses find that it is better to determine their margins by looking at what customers are likely willing to pay for specific goods or service and then working backward from there. Ideally, you will be able to reduce that price by five percent to ensure you are truly competitive in a cost-conscious world. While it might not seem like much, this extra five percent is extremely important as customers are constantly on the lookout for the next sale, regardless of how much is actually being saved. The mental benefits that come along with being five percent better than those around them will be more than enough for them to commit to your product or service over all the rest.

*Value add:* Regardless of what your business does, you will find that there are Lean principles that can be implemented in order to improve the overall amount of value you are providing for your customers while also showing them you appreciate their business and respect them as individuals. What's more, you will be able to address the potential for waste in your organization while also maintaining flow and work to achieve perfection.

Often, you can manage this by doing something as simple as listening to your customers' specific wants and needs which will make it easier for you to determine what they really value the most when it comes to the niche your company habits. Value is most often generated by adding on something tangible that either improves or modifies the most common aspects of the good or service being provided. The goal is that this improvement is something the customer is willing to pay for, so when they receive it for free, they see this as a viable reason for your service to cost more out of the gate. It is also very important that the added value is very easy for the customer to claim because otherwise, they will feel that you have misled them.

*Cost reduction:* As the Lean system is already quite big on cutting down on waste in all of its forms, it should come as no surprise that it has some ideas when it comes to cost-reducing measures. For starters, it is important to understand that when it comes to Lean, all the different types of waste can be broken down into three types.

Muri is the name for the waste that forms when there is too much variation within common processes. Muda is the name given to seven different types of waste including:

- Transportation waste is formed when parts, materials or information for a specific task are not available because the process for allocating resources for active products isn't where it should be.
- Waiting waste is created if a portion of the production chain has ideal time when they are not actively working on a task.
- Overproduction waste is common if the demand exceeds supply, and there is no plan in place to use this situation to the business's advantage. The Lean systems are designed to ensure that this number reaches zero so

supply and demand are always in balance with one another.

- Defective waste is known to appear when some segment of the standard operating process generates some issue that needs to be sorted at some point down the line.
- Inventory waste is known to appear if the production chain ends up remaining idle between runs because it doesn't have the physical materials needed to be constantly running.
- Movement waste is generated when required parts, materials or information needs to be moved around successfully to complete a specific step in the process.
- Additional processing waste is generated if work is completed that does not generate value or adds value for the company.
- A commonly added eighth muda is the waste created by the underutilization of your team. This can occur whenever any member of the team is placed in a position that doesn't utilize them to their full potential. It can also refer to the waste that occurs when team members have to perform tasks for which they have not been properly trained.

Muda also comes in three categories, the first of which is muda that doesn't directly add value but also cannot be easily removed if the system is going to continue working properly. When faced with this muda, the goal should be to work to minimize it slowly as a precursor to removing it completely. The second type of muda is that which has no real value, whatsoever, and you should work to remove it directly once you've become aware of it. Finally, the third type of muda doesn't directly add value but is required for regulatory purposes of one type or another. While it may be annoying, this type of muda is unavoidable in most instances and the best that you can do is ensure you are always updated to any relevant policies.

# Chapter 2: Creating a Lean System

## Lean leadership

With so much emphasis placed on improving efficiency, the Lean process naturally puts a lot of emphasis on team leaders who should be working hard to directly inspire their teams to adopt the Lean mindset. In the end, many Lean systems live and die by the leadership involved, which means it is important that those who are put in charge of leading the Lean transition are able to not only explain what's going on but are truly committed to the work that is being done as well. Some of the things that Lean leaders should strive to emphasize include:

*Customer retention:* When it comes to customer retention, Lean leaders need to take the time to consider not only what their customers want at the moment, but what they are likely to want in the future as well. Additionally, it is important to understand what a customer will accept, what they will enjoy, and what they will stop at nothing to obtain. The Lean leader should also work to truly understand the many ways the specific wants and needs of their target audience throughout the customer base.

*Team improvement:* In order to help their team members be their best, Lean leaders should always be available to help the team throughout the problem-solving process. At the same time, they are going to need to show restraint and refrain from going so far as to take control and just do things on their own. Their role in the process should be to focus on locating the required resources that allow the team to solve their own problems. Open-ended questions are a big part of this process as they will make it possible for the team to seek out a much wider variety of solutions.

*Incremental improvement:* One of the major duties of the Lean leader is to constantly evaluate different aspects of the team in order to ensure that it is operating at peak efficiency. The leader will also need to keep up to date on customer requirements, as this is something that is going to be constantly changing as well. Doing so is one of the only truly reliable ways of staying ahead of the curve by making it possible to streamline the overall direction of the company towards the processes that will achieve the best results.

In order to ensure that this is the case, the Lean leader will want to make time in their schedule to look at the results and then compare them to the costs as a means of discovering the best ways to use all the resources available to them at the given time. This will include things like evaluating the organization as a whole in hopes of making it more efficient and reliable. It will also involve evaluating the value stream to ensure that it satisfies the customer on both the macro and micro level.

*Focus on sustained improvement:* It is also the task of the Lean leader to ensure that improvements that are undertaken are seen through to the end as well. This will often include teaching the team members the correct Lean behaviors to use in a given situation and also approaching instances of failure as opportunities for improvement and innovation.

**Three actuals**

Lean leaders typically use a different leadership style than many of their peers, largely because being a Lean leader requires an understanding that the best way to analyze a situation is to physically be in the space where the situation occurred. Once there, the Lean leader needs to consider what is known as the three actuals, the broadest of which is known as Genchi and is the issue that led the leader to come to the place in question. Genbutsu is the idea that it is important to view what is being created or provided in action before making any moves. Finally, Genjitsu says it is always best to gather as much information as possible before making a decision one way or the other.

## Creating a Lean system

In order to create a Lean system that lasts, the first thing you will need to do is consider the absolute simplest means of getting your product or service out to the public and put that system into effect. From there, you will need to continuously monitor the processes you have put in place to support your business in order to ensure that improvement breakthroughs happen from time to time. The last step is to then implement any improvements as you come across. While there are plenty of theories and tools that can help you do go on from there, the fact of the matter is that creating a Lean system really is that simple. Many of the chapters in this book will consist of deep dives on various tools that will make this process as easy to set up and as most likely to stick as possible.

*There's more to business than profits:* When using the Lean system, the end goal is to determine the many ways that it might be possible to improve the efficiency of your business. While an increase in profits is often a natural result of this process, this should not be the primary motivating factor behind undertaking a Lean transformation. Instead, it is important to focus on streamlining as much as possible, regardless of what the upfront cost is going to be since you can confidentially expect every dollar you spend to come back to you in savings.

There are limits to this, of course, and at a certain point, the gains won't be worth the costs. To determine where this line is, you can use a simple value curve to determine how the changes will likely affect your bottom line. A value curve is often used to compare various products or services based on many relevant factors as well as the data on hand at the moment. In this instance, creating one to show the difference between a pre- and post-Lean state should make such decisions far easier to make.

*Treat tools as what they are:* When many new companies switch to a Lean style of doing things, they find it easy to slip into the trap of taking tools to the extreme, to the point that they follow them with near-religious fervor. It is important to keep in mind that the Lean principles are ultimately just guidelines and any Lean tools you use are just that, tools which are there to help your company work more effectively. This means that if they need to be tweaked to better serve your team and your customers, then there is nothing stopping you from doing just that. Your team should understand from the very beginning the limits and purpose of the Lean tools they are being provided and, most importantly, understand that they are not laws.

*Prepare to follow through:* Even if you bring in a trained professional to help your team over the initial Lean learning curve, it will still ultimately fall to you, as the team leader, to ensure that the learned practices don't fall by the wayside as time goes on. It takes time to take new ways of doing things and turn them into habits, and it will be your job to keep everyone on until everything clicks and they start operating via the new system without thinking about it. Likewise, it is important that you make it clear just why the Lean process is good for the team as a whole and for the individual team member as if they are personally invested in it, then it is far more likely that they will stick with it, even if the going gets tough.

# Chapter 3: Setting Lean Goals

In order to eventually make the right changes to your business, the first thing you need to do is ensure that you set the right goals. In order to make sure that your goals will put you on the right track, you need to ensure they are SMART which means they are specific, measurable, attainable, realistic and timely.

*Specific:* Good goals are specific which means you want to be sure that the goal you choose is extremely clear, especially when you are first starting out, as goals that are less well-defined are much easier to avoid doing in favor of activities that provide more positive stimulation in a shorter period of time. Keeping specific goals in mind will instead make it much easier for you to go ahead and power through whatever task you are currently undertaking.

When you aren't quite sure if the goal you have chosen is specific enough to actually improve your chance of changing for the better, you may be able to figure it out by running through the who, why, where, when, and how of the goal. Specifically, you are going to want to consider who is going to be involved with you when it comes to the completion of the goal? What exactly is it that is going to be accomplished? Where will it be taking place, why it is important that you ensure it is completed as quickly as possible, and how exactly you can expect to go about doing it. Once you can answer all five of the big questions, then you know you have a goal that is specific enough to generate the type of results that you are looking for.

*Measurable:* SMART goals are those which can be broken down into small, easily manageable chunks that can be tackled one piece at a time. A measurable goal should make it easy to determine when exactly you are headed off track so that you can self-correct as quickly as possible. Measuring your progress will make it easier for you to keep up the good work.

*Attainable:* Perhaps more important than anything else, if a goal that you set is unattainable, especially the first goal that you set using this system, then you are going to unknowingly be wasting valuable time and energy while creating negative patterns that end in failure. What's more, you will end up reinforcing fixed mindset ideals, making this a bad choice any way that you look at it. This means that when it comes to setting goals, you are going to want to have a clear understanding of the current situation and anything going on with the business that would make it less likely to succeed as far as that goal is concerned.

*Realistic:* A good goal is one that is realistic, in addition to being attainable, which means that you can expect success without something extremely unlikely being required to push reality into your favor. An ideal goal is one that is going to require a good amount of work to achieve, while still remaining not too difficult as to become unrealistic. Additionally, you are going to want to shy away from goals that you can meet without putting any real amount of effort as goals that are too easy can actually be demotivating as it then becomes easy to continue putting them off until they eventually fade into oblivion.

*Timely:* Studies show that the human mind is more likely to actively engage in problem-solving behaviors when there is a time limit involved in the successful completion of the task in question. What this means for the goals you are setting is that if you have a firm completion date in mind for when you want to have reached your goal, then you will work harder in the period leading up to that date. This means that you are going to want to pick completion dates that are strict enough to truly motivate you to do whatever it is you have in mind, while at the same time, not being so strict that there is no realistic way that you can complete the task on time. The goal here is to throw a little extra hustle into your step and not force you to keep a grueling schedule, thus, ensure that you can always meet the schedule you set for the best results.

## Policy Deployment

Also known by the name Hoshin Kanri, policy deployment is a way of ensuring any SMART goals that are set at the management level ultimately filter down to the rest of the team in a measurable way. Making proper use of policy deployment will essentially ensure that anything you are planning to put into effect doesn't accidentally end up creating more problems than it ultimately solves. It will also help to ensure as little waste as possible is generated as a result of things like inconsistent messaging from management or all around poor communication. The goal in this instance should not be to force various team members into acting in a specific way. It is about generating the type of vision for the business that everyone can appreciate and understand how it pertains to both the team and the customers.

*Implementing the plan:* Once all of the relevant SMART goals have been finalized, the next thing you will want to do is to group them together based on which members of the team will ultimately be tasked with solving them. Keep in mind that the fewer number of goals, the more likely it is that they will be acted upon in a reasonable timeframe. If your goals cannot be generalized in such a way, it is important to instead begin with the ones that are sure to make the biggest difference overall and work down the list from there.

Regardless of what goals you ultimately settle on, it is important that you take special care to ensure that there are no goals that do not have one person specifically assigned instructions to keep tabs on the overall progress while providing status reports when needed. This person should also be someone who can be counted on to make it clear to the other team members how important the goal is for the business as a whole and how it will make things easier in the long run.

*Consider your tactics:* Those who will be responsible for making the goal a reality should, in turn, be the ones who decide how the goal can best be completed by the team as a whole. However, this process should still include back and forth interaction between all levels of the team just to ensure that the tactics and the goal align properly. Tactics are likely to change as the goal heads towards success, which means it should be studied from time to time to ensure they remain appropriate for the goal in question.

*Moving forward:* Once the tactics have been agreed upon by all parties, it will then be time to actually put them into practice. This will be the stage where the team can really take over, though quality goals should still require buy-in from relevant parties. During this period it is important to ensure all communication from management is on message, to properly ensure that actions and broader goals will continue to align.

*Review from time to time:* It is important to keep in mind that once the action is in progress, the team leader will need to change the action as needed. This means that they will also be monitoring things as they proceed, hopefully, according to plan. Remember, Lean systems are always being improved upon, which means your goals and their implementation should be no different.

# Chapter 4: Simplifying Lean as Much as Possible

All of the products and services that are generated by your business have a mixture of three different value streams that can ultimately be used productively if you take the time to understand them fully. These include the concept to launch stream, the creation of customer stream and the order to customer stream. In order to ensure you are getting the greatest overall value out of all of the processes your business finishes, it is important to look at a value stream map as it is an excellent way to ensure you are maximizing efficiency at every turn.

The average value stream map will include everything that ultimately comes together to generate value for the customer including activities, people, materials, and information. To properly visualize a value stream, you will want to follow the Plan/Do/Study/Act process, also known as the Lean cycle. To get started, you will want to plan out the task ahead by focusing on one goal at a time. From there, you will want to make a list of everything that will need to be done in order to ensure the task is finished successfully. This is then followed by the step of following through, the results being studied, and acted upon as required.

**Create your own value stream map**

A properly constructed value stream map is a vital part of the process as it will allow you to see the big picture by mapping out the entire flow of resources from their disparate starting points all the way through when they come together and eventually make it into the hands of the customer. As such, it then makes it far more of a manageable task to determine the points in the process that are bottlenecking the overall efficiency of your business's process and thus, taking the first steps towards adopting Lean processes.

While one person can certainly work through the following steps, the value stream maps that prove to be the most effective are often those created by the entire team, so that those who are the most knowledgeable about each step will be able to give their two cents as well. Your initial value stream map should be thought of as a very rough draft and should be constructed as such, which means planning it out in pencil and expecting lots of rewriting as you go along.

*Consider the process:* The first step in this process will be to consider exactly what it is you will be mapping. For businesses that are first starting out with the Lean system, you will want to begin by considering the various processes that are ultimately going to prove to be of the greatest value to the team as a whole and then work down the list from there. If you still can't decide where to start, then you will want to turn to your customers, consider what they have to say and start with the areas where you regularly receive the most complaints.

What is known as a pareto analysis is an effective tool at this juncture as it can make it easier for you to find the right place to start if you aren't sure where your efforts will be best put to use. It is a statistical analysis technique that can prove especially useful if you are looking at a few different tasks that are sure to generate serious results if only you could decide which one to use first. The goal, in this case, is to focus on the

20 percent of your business that, if nurtured, could ultimately generate 80 percent of your total results. Your initial value stream map may focus on only a single service or product or on multiples that share a significant portion of the process.

*Choose your shorthand:* The symbols you use to denote various stages of the process you are mapping don't really have any hard and firm guidelines as they are going to be unique to every project and every business. Regardless of what you and your team ultimately choose, it is important to create a list of all of the symbols you are using and what they mean so that anyone who comes in after the fact can easily get caught up. From there, it is important to stick to the designated symbols and not make anything up on the fly. Additionally, if the business is working on more than one value stream map at a time, it is important that the symbols correspond between the two. Otherwise, things can quickly spiral into illegibility.

*Set limits:* If taken from a broad enough scope, virtually every value stream map for your business can be connected to other value stream maps or go into greater detail. At some point, however, this is going to be counterproductive and you will have to set limits on what the value stream map is going to account for if you ever hope to successfully move forward. Likewise, if you let this part of the process get out of hand, then the map can lose focus and become less useful as a result.

*Start with steps that are clearly defined:* After you have a clear beginning and end for the process you are mapping, the next thing to do is to make a list of all of the logical steps that need to be taken from start to finish. This shouldn't be an in-depth look at every link in the chain, but instead, should be an overview of the major stages that will need to be looked at more closely as the process moves towards completion.

*Consider the flow of information:* One important step in the value stream mapping process that sets it apart from other similar mapping processes is that each value stream map also accounts for the way that information flows throughout the process from beginning to end. What's more, it will also chart the way information passes between team members as well. You will also need to ensure it takes into account the ways the customer interacts with your business, in addition to how frequently such interactions occur. You will also need to ensure the communication chain includes any suppliers or any other third parties the company deals with.

*Further details:* When it comes to breaking the process down to its most granular level, you may want to include a flow chart with your value stream map as well. A flow chart is a great way to map out the innermost details of how a given process reaches completion. This is also an excellent way to determine the types of muda you are dealing with, so you can consider if they can be removed from the process.

If you are interested in considering the ways your team physically moves around your space, then a string diagram can also prove effective. To generate this type of diagram, you will map out your business's workspace by drawing in what each member of the team has to do and where they have to go in order to fully complete the process. You will want to draw different team members or different teams in different colors to keep things from getting too confusing. From there, charting the flow of information as it relates to this data can lead to surprising conclusions regarding flaws that might otherwise go unnoticed for years.

*Collecting data:* When it comes to outlining your initial map of a value stream, you may find yourself with certain aspects of the process that require additional data before anything can be determined with any real degree of certainty. The data that you may need to track down will include:

- cycle time
- total inventory on hand
- availability of the service
- transition time
- uptime
- number of shifts required to complete the process
- total available working time

When it comes to collecting this data, it is important to always remember to go to the source directly and find the details you are looking for rather than making assumptions. Furthermore, it is important to get the most updated numbers possible as opposed to looking at older, more readily available figures or hypothetical benchmarks. This may mean something as hands-on as physically keeping an eye on every part of the process in question so you can take relevant notes.

*Watch the inventory:* Even if you are relatively certain about any inventory requirements for the process in question, it is vital that you double check before you commit anything to the value stream map. Minor miscalculations at this point could dramatically skew your overall results and essentially nullify all of your hard work if you aren't careful. This means you definitely need to adopt a measure twice in order to see the best results. After all, inventory is prone to building up for a wide variety of reasons and there is a good chance that you won't know it until you take a closer look and do a once over on what's really on hand. You can also use this step as an excuse to take stock of exactly what the team is working with and determine how far it will actually stretch effectively.

*Using the data:* After you have finished visualizing the steps found in your most important process, you are now ready to use it as a means of determining where any problem points might be. You will especially want to keep an eye out for processes that include redoing any previously completed work, anything that requires an extended period of resetting before work can begin again, or long gaps where parts of the team can do nothing except wait for someone else to finish, those that take up more resources than your research indicates you should or even just those that seem to take longer than they should for no particular reason.

*Generate the ideal version of the value stream:* After you have determined where the bottlenecks are occurring, you will want to create an updated value stream map that represents how you want the process to proceed once you have everything properly sorted. This will provide you with an A to C scenario, where figuring out the pain points represents B. Ideally, it will also provide a clue as to how you can go about eliminating the waste from the process in order to create an idea, which you can really strive for both in the short and the long-term.

Once you have determined the ideal state for the process, you can work out a future value stream map that will serve as a plan on how to take the team from where you are currently to where you need to be. This type of plan is often broken down into sections that last a few months, depending on what needs to be done. Additionally, most future value stream maps will come with multiple iterations because they will need to change several times as the project nears completion.

When working through various variations of the value stream map, it is important to pay close attention to the lead time available for various processes. The lead time is the amount of time it will take to complete a given task in the process and, if not utilized as efficiently as possible, it can easily lead to a wealth of bottlenecks. Remember, when it comes to creating the best value stream map possible, no part of the process is beyond scrutiny.

# Chapter 5: Lean and Production

When putting together a Lean system for your business, you will eventually determine where the waste is hiding in your current processes, which is when it will be time to consider what can be done about the flow of the process. Often, the answer will come in either the form of a continuous flow model or a batch model.

*Continuous flow model:* The ideal version of the continuous flow model sees the customer order a product or service before the necessary steps are taken to generate the product or service that is being paid for. The product or service is then delivered to the customer who then pays for their order. The end result here is that there is no downtime between when the customer puts in their request and when it is completed. Furthermore, every step is going to smoothly flow into the next as a means of ensuring that overall downtime is reduced as much as possible.

The biggest upside to the continuous flow model is that it allows business to make assumptions and plan for the future based on a profit level that prioritizes continuity and stability. A continuous flow setup also features less waste than other types of processes. The biggest downside, however, is that this type of scenario can be difficult to produce reliably as every step in the process is rarely equal, regardless of how clear the value stream map might be. If you are striving to create a continuous process scenario, then you should be aware that new problems can also appear quickly if your available margin for error begins to shrink.

In order to persist despite these drawbacks, you will want to do your best to attack these problems head-on and be determined to push through them if you hope to find success. Additionally, if you hope to choose this route, it is important to start your journey to a Lean system with this in mind, as a continuous system is only going to work if every part of the system is completely in sync with all the rest.

Heijunka is a useful tool when it comes to facilitating this process as it promotes leveling out the quantity and quality of the process over a prolonged period of time in hopes of making everything as efficient as possible and, what's more, to expressly prevent batching. While it might sound complicated, in reality, this process can be as simple as making sure your team has all of the storage space they need to organize the various parts of the project. They store them in folders that are organized based both on the frequency of use as well as the due date. Folders that are currently in use can be stacked vertically on top of one another while those that are idle can be stored horizontally out of the way somewhere. There are also numerous other types of organizational methods that promote heijunka, so it can be helpful to explore them all to see what offers the most benefit to your business.

*Batch production:* Unlike with the continuous flow model, with the batch production model, the steps in the process to create the product or service are all completed in bulk one after the next. This makes it the superior choice when it comes to situations where what the process generates is evergreen as this will allow it to be stockpiled as a direct counter to erratic customer demand. Depending on the specifics of your business, batch production can also dramatically decrease the associated production costs as few team members can move from step to step instead of having all the steps operating at once. It also provides lots of opportunities when it comes to cross-training.

As a general rule, you can count on batch production to be less productive when there are a greater number of individual steps that are required to complete the process. This is because the starting and stopping times would need to be calculated for each which can add up quickly if batch sizes are quite large. This can also potentially create a delay if a customer places a large customized order when a batch is already in the middle of production.

## Takt time

Short for the word Taktzeit, Takt time is a variation of the Japanese principle of measuring time, despite its German name. Despite the fact that it is primarily used in production environments, it can have a beneficial effect on most tasks performed in a business environment as well. Specifically, Takt time is the time it takes for a team to start a new process after completely finishing the last, assuming the production rate is equal to the rate of customer demand.

*Determining takt time:* If your team completes processes one at a time throughout the workday, the takt time of that process can then be determined by taking the time that has elapsed between two processes, assuming of course that demand is still being met. This means it can be written as $T = Ta/D$. In this case, T is your Takt time, Ta is the amount of time available to finish processes, and D is the amount of demand that the process experiences.

You will not want to automatically take these results as fact, however, as it is rare to find a team that can run at peak efficiency at all times. As such, when it comes to determining an accurate takt time, you will want to add in some wiggle room here to compensate for the fact. From there, you will want to adjust your takt time based on additional customer requirements or team demands.

*Takt time benefits:* Once you have determined the accurate takt time for your business's processes, you will find that a number of additional beneficial options open up to you. First and foremost, you will find that it is clear which steps in the process are the bottlenecks which will make it easier to take steps to mitigate them specifically. Likewise, if you have any processes that typically go off the rails, that problem will be made apparent as well.

As a general rule, takt time places additional emphasis on steps that add value to the process as a whole, which makes it easier to use if you are looking for muda to remove from the process. What's more, once the team gets used to the concept of takt time, they will find that it is much easier to track how productive they are being throughout the day.

*Be aware:* Takt time is not a set-it and forget-it type of affair which means that if you find that your demand changes dramatically, then you will need to recalibrate all of your takt time to adjust for this fact. This also means that if your demand isn't relatively stable, then determining your takt time might not be very beneficial on its own. If you try and force your process into a takt timetable and it isn't a good fit, then all you will end up doing is causing even more waste in the long run.

Likewise, you will need to be aware of the way in which the products or services produced by your processes fit together or else, you risk creating bottlenecks anyway which will throw off the accuracy of your takt time. As a general rule, the shorter the takt time, the greater the amount of strain that resources including both machinery and people will experience on a regular basis.

# Chapter 6: Run a Lean Office

One of the truly great parts of the Lean system is the potential it holds when it comes to standardization, specifically when it comes to minimizing waste. Much like when it comes to setting goals, setting work parameters that are clearly standardized makes it easier to answer specific questions about the process. This should include things like who will follow through on the process once it has been outlined, how many people will it take, what will the end result be, what the metrics for success should be, what is required to meet them, how long the process will take and more. These are all questions that ultimately need to be asked in order to guarantee your standardization measures don't end up creating new problems instead of solving existing ones.

Workflow standardization is not expressly designed to ensure that processes are completed as quickly as possible. Rather, it is about utilizing the most effective practices possible in order to ensure they are completed with the same level of reliable quality each and every time. You will also do well to remember that standard practices will naturally change over time as improvements to safety, quality, and productivity are found. You will want to take care to avoid becoming so reliant on a single type of standardization that you end up actually allowing it to hold you back from future progress.

With that being said, it is also important to avoid falling into the trap of undertaking a round of standardization solely for standardization's sake. Instead, it is important to consider if standardization is really the right choice by considering the various processes already in place and asking yourself if they would be of a higher quality, performed to a higher safety standard or completed with less waste. If you move forward, and this is not the case, then all you are doing is inviting in waste.

Furthermore, standardization should involve more than simple instructional documents. It should be created from the input of those who perform the processes on the regular and then combined with a fresh round of customer feedback to ensure all bases are covered. The reasons for the standardization process should be clear to everyone involved before getting started for the best results.

## KPIs

KPIs, also known as key performance indicators, are extremely useful when it comes to determining the ideal steps to take during the standardization process. KPIs are also useful when it comes to measuring the overall success of the process as a whole based on numerous different metrics. Choosing the right KPI to focus on is a matter of considering what metrics you value most at the moment as well as in the long-term. There are a wide variety of indicators to choose from, all of which are useful in different circumstances and when it comes to accurately defining specific values. Essentially, each KPI can be considered an object which is useful in various value-add scenarios.

*Choosing indicators:* When it comes to identifying the KPIs you want to use for your business, the first thing you will need to do is ensure that your process is already well-defined as this will help you have a true handle on the specifics of every aspect of the process as well as the best ways to determine the ideal means of completion. It is important to only stick with indicators that are relevant to the goal you are currently working towards. Otherwise, they can easily be altered dramatically by factors that are literally outside of your control.

Much like with your goals, it is important that your KPIs are SMART and that they clearly indicate specific information for a specific purpose. You will also want to choose options that are easily measured while still providing accurate results if at all possible. Much like goals, KPIs are useless if they are not realistically achievable. The most effective KPIs are those that are relevant to the success of the business in the moment or in the future while also including an element of time that has specific periods as they relate to the data.

*Be aware:* It is also important to keep in mind that while determining specific KPIs isn't too difficult, keeping track and compiling the relevant data can be more difficult than it first appears. Furthermore, additional values, including those for things such as staff morale, are difficult to gauge accurately. Before you invest resources into generating KPIs, it is important to first make sure they are adequately measurable and useful. Otherwise, you will be on your way to creating even more waste.

You will also need to ensure the focus remains on keeping the KPIs on the data that they are detailing and use it as a means of determining the overall health of the business as opposed to a set of numbers that can only move one way. If your team ends up too focused on reaching a predetermined KPI, the data they return will be biased and inaccurate.

# Chapter 7: Kanban

Kanban is a method of scheduling that is often used once a Lean system has been put in place. It serves as a type of inventory management whose end goal is to minimize waste in the supply chain. It also tends to come in handy when it comes to pinpointing problems as it makes these problem areas stand out more than they otherwise would. You will also find it useful when it comes to locating the upper end of work related to inventory that is currently underway to ensure the process doesn't overload.

This is a demand-driven system which means, it is often implemented as a means of ensuring quicker turnaround times while at the same time limiting the required inventory and increasing the overall level of competitiveness between the implementation team. It was first put into effect by Toyota in the 1940s after the company performed a study on supermarkets and decided to use similar practices in order to keep their factories optimally stocked at all times. This eventually became one of the core Kanban ideas of keeping inventory amounts level with consumption rates. Additional supplies are then added based on a predetermined set of signals to ensure that stock remains near the ideal level at all times instead of dipping low or overflowing at certain points. The signals in question are all based upon customer signals which means they can change at the moment if needed.

### Kanban rules

- Each process creates an amount set by the

Kanban.

- Following processes collects the number of items set by the Kanban.
- Nothing is created or moved with a Kanban.
- Kanban is attached to related goods.
- Defective products are not counted in the Kanban.
- The fewer the Kanban, the more sensitive the system is.

*Kanban cards:* Kanban cards are the means by which signals are used to keep the entire team on track as it moves through the process. While they were actual cards when the system was created, these days, there is a wide variety of software out there that will provide the relevant signals without bringing physical cards into the process. Kanban cards generally represent consumption via a lack of cards in one area which, by necessity, drives another part of the process to do what needs to be done in order to pass the relevant cards along.

These days, the electronic Kanban system is even more effective than its physical predecessor, making it a sure thing to ensure that cards get where they need to be when they need to be there. These systems often mark set types of inventory with specific barcodes that are then scanned throughout the process. Each scan then sends a specific message to the Kanban program which routes it as needed.

*Kanban types:* There are two main varieties of Kanban systems namely production systems and transportation systems. Production systems are sent as a means of authorizing production or a specific number of items, while the transportation systems are used as a means of authorizing the movement of specific items once they have been created.

*Three bin system:* An example of a basic type of kanban system is the simple three bin system for the supplied parts in scenarios where manufacturing does not take place in-house. One bin represents the factory floor (or the primary point of demand anywhere else), the second bin represents the factory store (the control point for the inventory), and the final bin represents the supplier. The bins then can have removable cards containing relevant product details along with any other important information.

When the factory floor bin empties out because the relevant parts were all taken up by various parts of the manufacturing process, the empty bin, and thus its kanban card, are then returned to the factory store (also known as the inventory control point). The factory store then replaces the empty factory floor with the full factory store bin which also contains its own kanban card. The factory store then sends the empty bin and its related kanban card on to a supplier. This, in turn, causes the full product bin from the supplier to eventually replace the empty bin on the factory floor and the process starts all over again. Thus, the process never runs out of product. This could also be described as a closed loop, since it provides the exact amount required, with only one spare bin so there is never oversupply. This 'spare' bin allows for uncertainties in supply, use, and transport in the inventory system. A good kanban system calculates just enough kanban cards for each product. Most factories that use kanban use the colored board system

# Chapter 8: 5s

When it comes to determining what wasteful processes you are dealing with, it is important to ensure the work environment is in optimum shape for the best results. The 5S organizational methodology is one commonly used system based around a number of Japanese words that, when taken together, are first rate when it comes to improving efficiency and effectiveness by clearly identifying and storing items in their designated space each and every time.

The goal here is to allow for standardization across a variety of processes which will ultimately generate significant time savings in the long-term. The reason it is so effective is that each time the human eye tracks across a messy workspace, it takes a fraction of a second to locate what it is looking for and process everything around it. While this might not be much if it happens now and then, if it is happening constantly across an entire team, then it can add up to serious time loss when taken across the sum total of the process in question

*Sorting:* Sorting is all about doing what can be done in order to always keep the workplace clean of anything that isn't required. When sorting, it is important to organize the space in such a way that it removes anything that would create an obstacle towards the completion of the task at hand. You will want to ensure that process-critical items all have a unique space that is labeled as well as a space that is designated for those things that simply don't fit anywhere else. Moving forward, this will make it easier to keep the space free of new distractions. Nevertheless, it will still be important to encourage team members to prune their personal space regularly to keep new obstacles from popping up.

*Set in order:* When it comes to organizing the items in the workspace themselves, it is important to ensure all the items are organized in the order that they will most likely be used. While doing so, it is important to take care to ensure that everything required for the most common steps remains readily at hand to reduce movement waste as much as possible. Over time, keeping things in the same place will ensure that the process can be completed faster each time as muscle memory takes over, and team members are able to reach for things without looking for them.

It is important to keep an open mind during this step since ensuring that the workspace is set up in such a way that ease of workflow is promoted may require more than a simple organization, it may require a serious rework of existing facilities. Additionally, ensuring everything is arranged correctly will make it easier for you to create steps for each part of the process that anyone new to that part of the process can follow.

*Shine:* Keeping the workspace clean is an essential part of maintaining the most effective workspace possible. It is important to emphasize the importance of daily cleaning both for the overall efficiency boost and its ability to ensure that everything is where it is supposed to be so that there are no issues the next time they are needed. This will also provide an opportunity to have a regular maintenance if any is needed, which will serve to make the office a safer place for everyone. The end goal should be that any member of the team should be able to enter a new space and understand where the key items are located in less than five minutes.

*Standardize:* The standardize step is all about making sure the organizational process itself is organized in such a way that it can be applied throughout the entire business structure. This will make it easier to maintain order when things get hectic and also ensure that everyone can be held to the same reliable standard.

*Sustain:* Sustaining the process is vital as taking a week or more to properly get everything in order only to have it all fall apart six months later is going to accomplish nothing in the long-term. As such, it is important to ensure that the organization is a vital part of the DNA of the business in moving forward. If things are truly sustainable in this regard, then team members will be able to successfully move through the process without expressly being asked to. Unfortunately, you won't be able to expect this type of sustainability overnight. It will require plenty of training and an adoption of the idea as part of the business's culture.

*Great starter tool:* If your plan for your business is to transition to additional advanced Lean concepts over time, then 5S is a great way to start moving employees in that direction. It is especially effective with employees who are extremely stuck in their ways as, once they initially get on board, they will be hard pressed to deny the benefits in completion times that come with the improved organizational version. This, in turn, will make it easier for them to get on board with additional changes that may come in the future.

As a rule, when rolling out a new system like this, you can expect team members to only care about two things, the way the new system is going to affect them specifically and if the Lean process has actually seen results. This is also what makes 5S a great starting point as it has easily understandable answers for each that anyone can understand once they see the first workspace transformed for efficiency.

*Knowing if 5s is right for your business:* While 5S is a great choice for some businesses, it is not a one-size-fits-all solution, which means it is important to understand both of its strengths and its weaknesses when moving forward. Perhaps its biggest strength is that when implemented successfully, it is sure to help your team define their processes more easily while also helping them claim more ownership of the processes they are associated with as well. This extra structure also has the potential to lead to a much greater degree of personal responsibility among team members which will lead to a greater feeling of accountability throughout the process. When everything goes according to plan, this will then lead to further improved performance and better working conditions for everyone involved.

What's more, implementing 5S also has the potential to more likely make long-term employee contributions thanks to an internalized sense of improvement. Ideally, this will continue until the idea of continuous improvement becomes the order of the day. When done correctly, using 5S will also provide further insight into the realm of value analysis, equipment reliability, and work standardization.

On the other hand, the biggest weakness of 5S is that if it, and its purpose, are not communicated properly, then team members can make the mistake of seeing it as the end goal and not a means to an end. 5S should be the flagbearer for success to come in the future, not the sum total of a company's journey into Lean processes. Specifically, businesses whose movement is constrained significantly by external factors will have a hard time using 5S, and companies that currently have a storage problem would do well to solve it before attempting a 5S transition.

Additionally, just because 5S is a great fit for many companies doesn't mean that it will be the best choice for your team. This is especially true for smaller teams or for teams where team members wear many hats. Just because it is a popular way to implement Lean principles doesn't mean that it is going to be right for everyone. Moving ahead anyway and enforcing organization simply for the sake of organization won't do much of anything when it comes to generating real results. Instead, it will only generate new waste and it will only continue to do so before it is abandoned entirely.

This is especially true for businesses that run on a wide variety of human interaction, various management styles, and other management tools. However, when the various aspects work together properly, they will actually end up generating extra value for the customer which is a vital part of any successful business. If you blindly press forward with a 5S mentality, however, then it can become easy to lose sight of the outcome for the customer in pursuit of a perfect outcome or a perfect implementation of 5S principles.

Above all else, when implementing 5S, it is important that you stress to your team that 5S is something that should be part of the natural work routine and standard best practices, not an additional task to be done outside of daily work. The goal of 5S is to enhance the effectiveness of the workflow at every step in the process. Separating out the 5S into its own separate layer is the complete opposite of what the process stands for.

# Chapter 9: Six Sigma

Six Sigma is the shorthand name given to a system of measuring quality with a goal of getting as close to perfection as possible. A company operating in perfect synchronicity with Six Sigma would generate as few as 3.4 defects per million attempts at a given process. Zshift is the name given to the available deviations between a process that has been completed poorly and one that has been completed perfectly.

The standard Z-shift is one with a number of 4.5, while the ultimate value is a 6. Processes that have not been viewed through the Six Sigma lens typically earn around a 1.5.

*Zshift Levels:* A Six Sigma level of 1 means that your customers will get what they expect roughly 30 percent of the time. A Six Sigma level of 2 means that roughly 70 percent of the time, your customers will get what they expect. A Six Sigma level of 3 means that roughly 93 percent of the time, your customers will be satisfied. A Six Sigma level of 4 means that your customers will be satisfied more than 99 percent of the time. A Six Sigma level of 5 or 6 indicates a satisfaction percentage of even closer to 100 percent.

*Six Sigma Certification Levels:* Six Sigma is broken into numerous certification levels depending on the amount of knowledge the person in question has regarding the Six Sigma system. The executive level is made up of management team members who are in charge of actively setting up Six Sigma in your company. A Champion in Six Sigma is someone who can lead projects and be the voice of those projects specifically.

White belts are the rank-and-file workers; they have an understanding of Six Sigma, but it is limited. Yellow belts are active members on Six Sigma project teams who are allowed to determine improvements in some areas. Green belts are those who work with black belts on high-level projects while also running their own yellow belt projects. Black belts lead high-level projects while mentoring and supporting those at other tiers. Master black belts are those who are typically brought in specifically to implement Six Sigma and can mentor and teach anyone at any level.

*Implementation:* Giving your team a compelling reason to try Six Sigma is vital to the overall success of the process. In order to ensure that Six Sigma is properly implemented, it is important that you properly motivate your team by explaining how crucial the adoption of a new methodology really is. The most common choice in these situations is to create what is known as a burning platform scenario.

A burning platform is a motivational tactic wherein you explain that the situation the company now finds itself in is so dire (like standing on a burning platform) that only by implementing Six Sigma is there any chance of long-term survival for the company. Having stats that back up your assertions is helpful, though, if times aren't really so tough, a bit of exaggeration never hurt. Adapting to Six Sigma can be difficult, especially for older employees and a little external motivation can make the change more palatable.

*Ensure the tools for self-improvement are readily available:* Once the initial round of training regarding Six Sigma has been completed, it is important that you have a strong mentorship program in play while also making additional refresher materials readily available to those who need them. The worst thing that can happen at this point is for a team member who is confused about one of the finer points of Six Sigma to try and find additional answers only to be rebuffed due to lack of resources.

Not only will they walk away still confused, but they will also be rebuffed for trying and not rewarded for taking an interest in the subject matter. A team member who cannot easily find answers to their questions is a team member who will not follow Six Sigma processes when it really counts.

*Key principles:* Lean Six Sigma works based on the common acceptance of five laws. The first is the law of the market which means that the customer needs to be considered first before any decision is made. The second is the law of flexibility wherein the best processes are those that can be used for the greatest number of disparate functions. The third is the law of focus which states that a business should only focus on the problem the business is having as opposed to the business itself. The fourth is the law of velocity which says that the greater the number of steps in a process, the less efficient it is. Finally, the last is the law of complexity which says that simpler processes are always superior to more complicated ones.

*Choosing the best process:* When it comes to deciding what process to apply the Six Sigma treatment to, the best place to start is with any processes that are already defective and need work to reduce the number of times they occur. From there, it will simply be a matter of looking for instances where takt time is out of whack before looking into those steps where the number of available resources can be reduced as well.

*Methodologies:* There are two main ways to get the most out of Six Sigma, DMADV and DMAIC.

DMAIC is an acronym that is useful when it comes to remembering five phases that can be useful when it comes to creating new processes.

- Define what the process should do based on customer input.
- Measure the parameters that the process will adhere to and ensure it is being created properly by gathering relative information.
- Analyze the information you have gathered.
- Improve the process based on the analysis you have completed.
- Control the process as much as you can by finding ways to reliably decrease the appearance of delinquent variations.

DMADV, on the other hand, also has five phases that correspond to the DMAIC phases.

- Define the solutions the process should be providing.
- Measure the specifics of the process to determine its parameters.
- Analyze the data you have collected up to this point.
- Design the new process using your analysis.
- Verify the results as needed.

*Deciding if Six Sigma is the right choice:* While the Six Sigma system has something to offer teams of all shapes and sizes, that doesn't mean that it is going to be the best fit for all of them. This is especially true as implementing it successfully depends on numerous different specifics, starting with the conviction of those who are looking to implement the system in the first place as well as the company's overall culture. This is why it is best to start with something less high-impact like 5S as a way to ease your team into things that are more of an overall change like Six Sigma.

When deciding if a Six Sigma transition is feasible, it is important to ensure that it is not seen as a fad and can instead be seen as an evolution of the ideals already in place. Generally speaking, the more involved the team leadership is from the beginning, the more onboard the rest of the team will be as well. It is important that the company culture is perceived to be one that supports this sort of positive change and to remember that if the management team can't come to a consensus on the new program, then it is sure to be dead in the water. This doesn't mean that absolutely every member of the team needs to be committed to the idea of Six Sigma from the start, but it does mean that the change needs to be institutional so the public front always needs to appear united.

After all, Six Sigma was founded on the idea of leaders mentoring those beneath them in order to ensure Six Sigma works as it should, but this need to be a full-time job for some people, at least until the new habits start to solidify among the team as a whole. Once this occurs, you can then count on the team members to keep one another on track. To ensure they get to this point, you are going to want to let them know that their progress is being tracked so that every team member constantly feels as though they are improving up until the point where they internalize the Six Sigma principles.

# Conclusion

Thank you for making it through to the end of *Lean Enterprise: The Complete Step-by-Step Startup Guide to Building a Lean Business Using Six Sigma, Kanban & 5s Methodologies.* Let's hope it was informative and able to provide you with all of the tools you need to achieve your Lean implementation goals. Just because you've finished this book doesn't mean there is nothing left to learn on the topic. Expanding your horizons is the only way to find the mastery you seek.

When it comes to implementing Lean techniques successfully, it is important to be realistic when it comes to the timeframe required to not just ensure the entire team is up to speed, but that they have internalized the core Lean principles you are trying to instill. You will need to take a long hard look at your team and your business as a whole and decide where the most work is going to need to take place. Every business has limited resources, after all. It is important to think wisely prior to allocating them.

While you can easily get sucked into a pattern of changing everything, in order to ensure your business really is as Lean as possible, you should keep in mind that discretion is the better part of valor and you should be sure to start by focusing on those things that will end up doing the most good before moving on from there. Don't forget, change for the sake of change won't do anyone any good and will likely serve to create more waste than it will eliminate. Ultimately, it is important to remember that creating a Lean business is a marathon, not a sprint, which means slow and steady wins the race.

Finally, if you found this book useful in any way, a review on Amazon is always appreciated!

Printed in Great Britain
by Amazon